Ten Years To The Singularity If We Really Really Try

… and other Essays on AGI and its Implications

Ben Goertzel

ISBN-10:1505550823
ISBN-13:9781505550825

Table of Contents

Preface

I've been thinking about Artificial General Intelligence (AGI) and the future of human and transhuman intelligence since I was 4 or 5 years old and first encountered such ideas in SF novels. As time went on my thinking veered from the purely speculative and futuristic to the practical and engineering-oriented; in the last two decades the pursuit of real-world AGI systems in the here and now has been the central theme of my career (most recently largely via the OpenCog project). But I've never taken my eye off the broader potential AGI holds, and its connection to various other aspects of current and future life – i.e. to almost anything you can think of!

I thoroughly believe that the transition from human intelligence to transhuman artificial intelligence is going to be far huger than any transformation in the history of humanity – more akin to the transition from bacteria to humans, or from rocks to bacteria, if one wants to draw historical analogies. In the shorter term, there will be plenty of dramatic implications for human society and psychology. And in the longer term, the implications will go way, way beyond anything we can imagine. Both the shorter-term and longer-term implications are amazing and exciting to think about.

During the period 2009-2011, when I had a bit more time for writing than I do now, I wrote a series of essays for H+ Magazine – the online zine I co-edit -- on the future of AGI, the Singularity and related topics. While these don't exactly flow together like chapters of a single-focused book, they do preoccupy with a common set of themes and present a coherent point of view. So I have gathered them here together in a single volume, along with a handful of non-H+-Mag essays from the same period – a couple that were published elsewhere, and a couple that languished on my hard drive unpublished till now.

Worth noting, perhaps, is that one thing this book DOESN'T contain is a detailed overview of my own work on AGI, which I've described in a variety of technical works before, and which is currently centered on the OpenCog open source AGI platform. I'm currently (mid-2014) cooking a non-technical book called "Faster Than You Think" which will cover this ground. Various chapters here in "Ten Years..." do touch on aspects of my own work, but only here and there in passing – not in a systematic way. What this book DOES give, though, is a broad overview of the way I think about AGI in terms of its relationship with other futurist technologies, and the amazing future that AGI and these technologies together are likely to bring to future humans and the future successors of humanity.

None of the essays collected here are technical. Some are very lightweight, at the level of a newspaper article. Some are more in-depth than that, at the level of a fairly intense pop-sci book or an article in a popular science magazine. It should all be readable by anyone with a couple years of university education in any field.

Some of the material here is "dated" in some respects – relative to mid-2014 when I'm writing this preface -- but I haven't made any effort to update it. By the time you're reading this, 2014 may be long gone as well. The essays collected here make some points that transcend their time of writing; and they also may have some value as a record of how their topics were being thought about during the particular period 2009-2011. AGI is advancing rapidly; and thinking about the Singularity and the future of humanity and transhumanity is also advancing rapidly. Humanity's thoughts about the (conceptually or calendrically) far-off future, written at a certain time, are generally at least as informative about the state of humanity at the time of writing, as they are about the far-off future.

Friedrich Nietzsche, in a draft preface for a book he planned to write with the title *The Will to Power* (different from the collection of notes that was released posthumously under that title), described the book as

"A book for thinking, nothing else; it belongs to those for whom thinking is a delight, nothing else."

The thoughts in this book are offered in a similar spirit. There is nothing definitive here. It's all necessarily quite preliminary, because we are still in the early stages of creating AGI; and even though we may be fairly close to a Technological Singularity in terms of calendar years, we have loads of technological, scientific and conceptual progress to pass through before we get there. All the matters discussed here are going to seem quite different when we get to Singularity Eve. But thinking through these topics, now – as I did in writing these essays; and as I invite you to do while reading them -- is a critical part of the process of getting there.

--

The cover art for the book was done by Zarko Paunovic.

1

Next Steps For – And Beyond – Humanity

Some introductory musings…

We humans have come such a long way in such a short time! Yet, we're just at the beginning of our evolution, with extraordinarily dramatic transformations soon to come.

According to our best understanding, the universe has existed for around 14 billion years, Earth for around 4 billion, multicellular life for around 2 billion years, and humans with roughly modern-style brains and bodies only couple million years or so.

Human civilization dates back only 10,000 year or so, and a lot more recently on most of the planet. In our relatively short time of existence, we humans have invented/discovered an amazing variety of stuff like language, mathematics, science, religion, fashion, sports, social networking, romance, art, music, corporations, computers, spacecraft – well, you get the idea.

The pace of our invention and discovery has been accelerating fairly consistently. There was a lot more new creation in the human world from 1000-2000 A.D. than from 0-1000 A.D.; and a lot more from 0-1000 A.D. than from 1000 B.C.-0 A.D., and so forth.

Certainly history has had its ups and downs, and the rate of human progress along various axes is difficult to quantify. But qualitatively, the accelerating rate of progress is plain to see. I've even felt it in my own lifetime; The pace of new gadgets available and the pace of scientific discoveries is now is incredible compared to when I was a kid in the 1970s.

However, all the change we've seen so far is trivial compared to what's coming. So far, the advances the human race has made, and the transformations it's undergone, have all been within the scope of the human. We've created new methods for communicating, new physical tools for doing stuff, and various techniques for modifying our minds and bodies. Our current lifestyles and thought-systems are extremely different from those of our hunter-gatherer forebears; indeed, the life and mind-state of a young, wired, fervently multitasking Silicon Valley or Hong Kong technology entrepreneur today bears scant resemblance to anything that existed 50 or 100 years ago. But, as dramatic as all this has been, there have been certain limitations. So far, the human body and brain have remained mostly invariant, and have provided constraint, structure and guidance to the unfolding changes.

The nature of the next step in humanity's progress has been apparent to visionaries for some time, and is lately coming to the attention of a larger swath of the population. We have created tools to carry out much of the practical work previously done by human bodies. Next we will create tools to carry out the work currently done by human minds. We will create powerful robots and artificially intelligent software programs – not merely "narrow AI" programs carrying out specific tasks, but AGIs, artificial general intelligences capable of coping with unpredictable situations in intelligent and creative ways. We will do this for the same reason we created hand-axes, hammers, factories, cars, antibiotics and computers – because we seek to make our lives easier, more entertaining, and more interesting. Nations and corporations will underwrite AGI research and development in order to gain economic advantage over competitors – this happens to a limited extent now, but will become far more dramatic once AGI technology has advanced a little further. The result will be something unprecedented in human history; At a certain point, we will no longer be the most generally intelligent creatures on the planet.

What will be the next step be beyond this? Where will the acceleration of technology ultimately lead us? Of course it's impossible for us to say, at this point. By analogy, I like to imagine the first humans to create a language with complex sentences, sitting around the campfire speculating about where the wild new invention of "language" is ultimately going to lead them. They might have some interesting insights, but would they foresee mathematics, Dostoevsky, hip-hop, PhotoShop, supersymmetry, remote-piloted cruise missiles, World of Warcraft or the Internet?

However, basic logic lets us draw a few conclusions about the nature of a world where powerful AGIs exist. One is that, if humans can create AGIs more intelligent than themselves, most likely these first-generation AGIs will be able to create AGIs with yet greater intelligence, and so on. These second-generation AGIs will be able to create yet smarter AGIs. This is what mathematician I.J. Good, back in the 1960s, called the "intelligence explosion." The dramatic potential consequences of this sort of intelligence explosion led science fiction writer Vernor Vinge, in the 1980s, to speak of a coming "technological Singularity."

Mathematically, one interpretation of the term "Singularity" would be a point at which some curve or surface changes infinitely fast. Of course, the rate of technological change is unlikely to actually approach infinity. Though the physical constraints we now perceive may well be surmountable, there are probably other limitations beyond our current understanding. But even so, the intelligence explosion may well bring a "Singularity" in the sense of a dramatic, incredibly hard to predict qualitative change in the world we live in.

A lot has been written about the Singularity and accelerating technological change in recent years. Famed inventor Ray Kurzweil, in particular, has done a huge amount to bring the Singularity meme to the world's attention, via his 2006 book *The Singularity Is Near* and his numerous speeches and articles. (I was pleased to have a brief appearance in the Ray Kurzweil documentary *Transcendent Man* – if you haven't seen it, check it out online!) I won't try to repeat here what has already been said by Ray, and others before him such as Damien Broderick in his excellent 1997 book *The Spike* -- not as catchy a term as "Singularity", but the same basic idea. I encourage you to check out Kurzweil and Broderick's books, if you haven't already, for an in-depth analysis of the human race's accelerating progress over the last decades, centuries and millennia.

Not everyone believes the Singularity is near, of course. On Kurzweil's kurzweilai.net website, you can also find counter-arguments by Kurzweil to many of the complaints made by detractors of the Singularity idea. Some thinkers whom I much respect think the Singularity idea is a bit overstated – transhumanist thinker Max More, for example, argues that we're more likely to see a fairly gradual Surge than a Singularity. Philosopher David Chalmers has laid out a rigorous analytical argument explaining why a Singularity is very likely to happen, following the basic logic of I.J. Good's "intelligence explosion" idea. But Chalmers is a bit more conservative than Kurzweil in the particulars. Kurzweil projects a Singularity sometime around 2045, but the Singularity hypothesis as Chalmers analyzes it has to do with a Singularity sometime within the next few hundred years, characterized by an explosion from human-level AGI to massively superintelligent AGI during a period of decades. I respect the care and rigor of Chalmers' analytical approach, but in the end I suspect Kurzweil is probably close to the mark. I even think we could see a Singularity well before Kurzweil's projected 2045, if the right alignment of scientific progress and economic interest comes about. This is one of the things I'm eager to tell you about in this book.

My own particular focus, as a scientist, is on the AGI aspect. The Singularity (or Surge, or whatever) is going to involve a variety of different technologies, including genetic engineering, nanotech, novel computing hardware and maybe quantum computing, robotics, brain computer interfacing, and a lot more. No doubt there will be new categories of technology that we can't now envision. However, in the "intelligence explosion" perspective, AGI plays a special role – it's the main technology catalyzing the next wave of radical change, taking us from the state of "humans with advanced tools but old fashioned bodies and brains" to a new condition that includes radically posthuman features.

The main focus of my own research is on the quest to create AGI with general intelligence at the human level, and with the capability to take the next step and create new AGIs with even greater general intelligence. Furthermore, I want to do this in a way that's going to have beneficial effects, for human beings in their current and future form and for the new intelligences we create. In addition to this, I also work on applications of simpler AI technologies to practical problems, such as the genetics of human longevity, and the analysis of financial markets. And I've put a fair bit of energy into organizing the AGI research community, such as launching a series of annual AGI conferences, and forming a professional research association called the AGI Society. I've dabbled with leadership roles in the futurist, transhumanist community, including a period as Chairman of Humanity+, which is the only international broad-based nonprofit organization devoted to technology and the future of humanity. All these experiences have given me a unique perspective on the present and future of the AGI field, which is what I'll try to convey to you in these pages.

Unless you're a very rare reader, many of the ideas I'll discuss in these pages will strike you as radical and "out there." But I don't mind coming across that way. I don't aim to shock, but I also don't aim to shape my message to suit currently culturally prevalent ideas. In this book I just call it like I see it.

2

The Top Priority for Mankind

In early 2009 I was contacted by some folks associated with the World Economic Forum – best known for their annual conference in Davos, Switzerland – to write an article for distribution at the Summer Davos World Economic Forum that summer in Dalian, China. The attendees at Davos are the world's political and business movers and shakers – top politicians and CEOs and philanthropists and so forth. My contribution was to be included (in Chinese translation) in a special issue of Green Herald Magazine, comprised of articles on the theme of "Mankind's Top Priority."

I like to get right to the point, so the title of my article was: "Mankind's Top Priority Should Be the Creation of Beneficent AI With Greater than Human Intelligence." The Summer Davos Forum is also called the "Annual Meeting of the New Champions", so I thought this was particularly appropriate. The "New Champions" phrase was presumably chosen to refer to the leaders of China, India and other emerging markets. But I wanted to question the assumption that the new champions leading humanity onwards would always continue to be humans.

Here is a section of what I wrote for the Davos participants:

> What should mankind's top priority be going forwards?
>
> This may seem to be a hard question, because there are so many different problems in the world, and also so many different opportunities for positive development.
>
> But the answer I'll propose is a very simple one. Our top priority should be the creation of beneficent artificial minds

with greater than human intelligence, which can then work together with us to solve the other hard problems and explore various positive avenues.

The pursuit of Artificial General Intelligence (AGI) is currently a small niche of the research and development world – most of the AI research done today focuses on the creation of highly specialized "narrow AI" systems solving isolated problems. But my contention is that AGI should receive dramatically more emphasis and should in fact be the human race's top priority going forward. The reasoning behind this contention is simple: whatever the hard problem at hand, a greater than human intelligence with the proper motivational system is going to solve it better than humans can.

This may seem a farfetched notion, but an increasing number of technologists believe that greater than human AI could arrive sooner than most people think, quite plausibly within the next 2-3 decades. Toward this end each year a number of researchers have gathered in an international conference on Artificial General Intelligence. If these optimistic predictions are accurate, clearly this has extremely important implications for the way we think about ourselves and our future.

The Benefits Advanced AGI Will Bring

The potential benefits to be delivered by advanced AGI are difficult to overestimate. Once we are able to create artificial minds vastly more intelligent than our own, there are no clear limits to what we may be able to achieve, working together with our creations. The possibilities quickly start to sound science fictional – but in the era of space travel, the Internet, genetic engineering, nanotech and quantum computing, the boundary between science fiction and reality progressively blurs.

Acceleration of Biomedical Research

According to contemporary biology, the human body is in large part a complex piece of molecular machinery, and once we have understood its operation sufficiently well, there is no reason in principle we shouldn't be able to repair its problems. The main limiting factor here is the human mind's ability to understand the biological data already collected, and to design new instrumentation able to gather more accurate data. The advent of AGI biologists should lead to radical improvements in medicine, including potentially an end to most or all human disease.

We can already see a start in this direction in the form of recent work applying AI to bioinformatics (such as the work of my own firm Biomind LLC) and biomedical text processing (see the MEDSCAN product by Ariadne Genomics, for instance). For example, in my own work, we used AI technology to discover the roots of Parkinson's Disease in mutations of mitochondrial DNA in brain cells – work which is now being commercialized as a diagnostic test, and being used to guide research toward a cure. When diseases like cancer and aging are finally cured, the odds are high that either narrow AI or AGI will play a major role – and there can be little doubt that AGI would get us there faster.

Discovery and Advancement of Alternative Energy Technologies

Affordable and sustainable energy is one of the most daunting problems facing humanity today. However, the brute fact is that, at the present time, by far the most affordable energy sources are fossil fuels (with limited roles for nuclear, geothermal, hydroelectric and other energy sources in particular situations). Technologies are known that are in-principle capable of providing vastly more efficient energy generation without consuming nonrenewable resources – but the process of refining these technologies to the point of economic practicality has proved a slow one, both because of funding limitations, and because of the fundamental complexities of the physical processes involved. It is easy to imagine that an artificial mind with modestly greater than human intelligence could discover the electrochemical basis for more effective solar panels, for example – thus revolutionizing the energy field.

Practical Drexlerian Nanotechnology

Back in the 1980s, in his path-breaking book Nanosystems, Eric Drexler proposed the creation of molecular assemblers – machines capable of piecing together structures out of molecules, into essentially any configuration desired... Much like a Lego or Erector set, but using real molecules. Today, nanotechnology is a thriving field of engineering, but most of the current work avoids the deeper aspects of nanotech that Drexler envisioned, and focusing instead on such valuable but narrower problems as using nanotech to create stronger materials, more efficient conductors, or fabrics with useful properties.

There is little doubt that Drexlerian nanotech will come; and firms such as Drexler's NanoRex, with their nanoscale Computer-Aided Design software, are pushing hard in that direction. But, one of the major challenges keeping us from getting there is that human intuition is simply not well-tuned for the nanoworld. AGI systems need not suffer from this limitation!

Human Cognitive Enhancement

If smarter-than-human AGI systems can revolutionize the world, why not smarter-than-human humans? Why not use advanced technology to make human brains better?

Indeed a number of neurobiologists are thinking hard about this problem. And at the present time, the worst bottleneck in this research direction seems to be the understanding of how the human brain works in the first place. In order to connect external devices to neurons in a useful way, we need to understand what the signals passed along those neurons mean; and currently our understanding of human brain dynamics is extremely crude. The solution to this problem has two aspects: the creation of better brain-imaging devices for noninvasively capturing highly spatiotemporally accurate neurological data, and the effective analysis of the data emanating from these brain-imaging devices. It is very clear that advanced AI technology would accelerate our understanding in both these areas. Making smart AIs may be the most effective path toward making smarter people.

Acceleration of AGI Research

Last but not least, one of the most promising application areas for a human-level AGI is AGI itself. As every software engineer knows, the design and implementation of complex software is a process that constantly *pushes against the limitations of the human brain – such as our limited short-term memory capacity, which doesn't allow us to simultaneously manage the states of more than a few dozen variables or software objects. There seems little doubt that a human-level AGI, once trained in computer science, would be able to analyze and refine its own underlying algorithms with a far greater effectiveness than any human being*

My screed on the amazing potential of AGI to solve all the world's problems surely struck the Davos crowd as a little far out there, compared to their usual fare of international politics, journalism, law, philanthropy and so forth. However, Davos is supposed to be about thinking big, and I wanted to encourage the participants to think *really* big. The creation of AGI with general intelligence at the human level or beyond will have a more dramatic, important and fascinating impact than anything on the program at Davos that year, or any other year so far.

3

Why AGI?

Before going any further, I suppose I'd better deal with the "AGI" issue. Everyone knows what "AI" is, so why do I insist on using the less familiar term "AGI"?

This tweak of terminology points at deeper issues.

The basic reason for talking about AGI is: Modern AI has a split personality.

In movies and SF novels, "AI" refers to robots or computer programs with a high level of intelligence and autonomy – R2D2 or C3PO or Hal 9000 or the Terminator or whatever.

However, in university computer science departments and industry research labs, AI usually refers to something far more prosaic – mainly the creation of highly specialized software carrying out particular functions that, when humans do them, are generally considered to require intelligence. The machine at the paint store that mixes your colors for you and mixes them in with the white paint, for instance, is an AI at work that has become so pervasive, we don't even think about the AI in use when we're at the paint store. The term "AI" needs to be used exclusively when we are referring to these very specific types of "intelligence" programs, for they are a far cry from an "AGI;" an actual "thinking machine," capable of discerning information coming in simultaneously from a number of varied and specialized sources.

So far, the only "AGI" in existence is the human brain. To many, the promise of building a software-based thinking machine requires the reverse engineering of our real brains, in an attempt to set up some kind of real time brain simulation, before we start feeding questions into it. I do not believe this is the proper approach. Reverse engineering the brain is a process that will take 100 years on its own to complete. We should be able to write software that thinks far before that time, as long as we're not hung up on its processes mimicking that of our real brains.

A new university student, head full of science fiction, might imagine the community of AI scientists around the world is working day and night trying to create computers smarter than people— trying to work toward computers capable of holding intelligent conversations, outsmarting Nobel Prize winners, writing beautiful poetry, proving amazing new math theorems. But a glance at the tables of contents of the leading academic AI journals or conferences would rapidly dash this student's hope, for instead they would find dry formal analyses of very particular aspects of intelligence, or applications of AI technology to perform simple puzzles or very particular commercial problems. AI research as it currently exists focuses almost entirely on what Kurzweil calls "narrow AI" – highly specialized problem-solving programs, constituting at most small aspects of intelligence and involving little or no spontaneity or creativity.

This split personality was not there at the inception of the AI field, back in the 1950s and 60s; it's something that evolved over time. The founders of AI were thinking about AI in the ambitious sense. Listen to Nils Nilsson, one of the early leaders of the AI field back in the 1960s, who reiterated his vision of AI forcefully in a 2005 article titled "Human-Level Artificial Intelligence? Be Serious!"

> *"I claim achieving real Human-Level artificial intelligence would necessarily imply that most of the tasks that humans perform for pay could be automated. Rather than work toward this goal of automation by building special-purpose systems, I argue for the development of general-purpose, educable systems that can learn and be taught to perform any of the thousands of jobs that humans can perform. Joining others who have made similar proposals, I advocate beginning with a system that has minimal, although*

extensive, built-in capabilities. These would have to include the ability to improve through learning along with many other abilities."

Hear, hear! This is AI in the grand sense – AGI as I conceive it.

The early AI pioneers had some brave, exciting ideas, and some rather deep insights. But then, as their grand goals proved unexpectedly tough to achieve, the field drifted away from its original goals.

Make no mistake, this focus on narrow problem-solving methods has yielded some exciting fruit. Practical, real-world narrow-AI work has yielded far more results than –those outside the AI field realize. It has led to a variety of valuable technologies, pervading nearly every area of society, e.g.

- the language processing AI underlying search engines like Google
- the planning and scheduling AI used extensively in government and industry, including many modern military operations
- grandmaster-beating chess programs
- industrial robots
- video game AI characters, playing a central role in games played by
- hundreds of millions of people
- AI financial trading systems
- AI data mining systems helping businesses, scientists and others with historical analysis and decision making (aka "business intelligence")

However, these narrow-AI programs have not progressed that far toward the original goal of the AI field, the construction of what I call AGI.

Only in the last decade has a significant subset of the field began to shift back toward the field's original focus, on creating AI systems with general intelligence at the human level and ultimately beyond. The term "AGI" arose as part of an effort to accelerate and deepen this shift of focus.

Terminology is not a terribly deep topic compared to the nature of intelligence or the construction of thinking machines, yet it has more influence on the development of science and engineering than one might think. "Black holes" have attracted a lot more research, as well as media attention, than would have been the case if they'd been called "gravitational sinks" (a more literal and transparent name). The field of "artificial life" flourished dramatically for a while, in part because of the funky animations it led to, and in part because of the sexy name. "Chaos theory" sounds way whizzier than "nonlinear dynamics," so the former is commonly used even when the nonlinear dynamics in question aren't technically "chaotic."

On the other hand, some fields like "bioinformatics" and "functional genomics" have extremely dry and boring names that conceal the dramatic wonder and importance of the subject matter. "Data mining" was fashionable for a while because of its exciting sound, but then it fell out of favor due to a period in which the technology was used crudely and led to a lot of meaningless results, and so now it's more suave to refer to "applied machine learning." And of course, both "data mining" and "machine learning" comprise topics typically taught in the "AI" courses in university – but it's sometimes considered preferable in industry not to refer to them as "AI" because the latter carries too much science-fictional baggage... Even though 98% of AI research done in academia and industry is quite prosaic with no direct relation to AGI. Even scientists can be influenced very substantially by the names used to refer to things.

The strong points of the term “AGI” are the obvious connection with the recognized term “AI,” and the connection with the “g factor” of general intelligence, well known in the psychology field. (It's what IQ tests are supposed to measure.) There are also some weak sides to the term though; Actually three weak points: “artificial,” “general,” and “intelligence!”

“Artificial” isn’t really appropriate – since AGI isn’t just about building systems to be our tools or “artifices.” I guess it’s about using artifice to build general intelligences though!

“General” is problematic, because no real-world intelligence will ever be totally general. Every real-world system will have some limitations, and will be better at solving some kinds of problems than others.

“Intelligence,” finally, is a rather poorly defined term, and, except in highly abstract mathematical contexts, fairly divorced from real-world systems. Nobody has a totally clear idea what it means.

However, in spite of these shortcomings, I actually like the term “AGI” well enough. It seems to be catching on reasonably well, both in the scientific community and in the futurist media. So Long Live AGI!

Why the Time is Ripe for AGI

But why am I so confident that building a powerful AGI is really feasible?

Doesn’t it worry me that the bulk of my fellow humans, and even the bulk of professional AI researchers, hold a different opinion from me on the feasibility of creating AGI in the near term?

In the big picture, it definitely doesn't bother me at all. The vast majority of major innovations in the history of humanity have been pooh-poohed by average people and experts alike. Of course, the vast majority of visionaries moving forward according to their wild-eyed ideas, while ignoring the naysayers, have proved them wrong after all.History provides a huge diversity of examples and situations, leaving one with no realistic choice but to evaluate each new situation on its own merits. AGI is something I've thought about for a long time, including reflecting on the fact that many others in the past have had misplaced confidence about the proximity of human-level AGI, and, by this point, I have a lot of trust in my own intuition on the matter.

Put simply there are five big reasons why I think advanced AGI is just around the corner:

1. **Computers and computer networks are very powerful** now, and getting more and more so.
2. **Computer science has advanced greatly**, providing all manner of wonderful algorithms, many wrapped up in easily accessible code libraries (like STL, Boost and GSL, to name three libraries I use in my own work), and many already engineered so as to be capable of running on distributed networks of multiprocessor machines.
3. **Robots and virtual worlds have matured significantly**, making it feasible to interface AI systems with complex environments at relatively low cost.
4. **Cognitive science has advanced tremendously**, so that we now have a pretty strong basic understanding of what the different parts of the human mind are, what sorts of things they do, and how they work together (even though we still don't understand that much about the internal dynamics of the various parts of the mind, nor how they're implemented in the brain)
5. **The Internet provides fantastic tools for collaboration** – both in terms of sharing of ideas among

groups of people (email lists, wikis, Web repositories of research papers), and in terms of collaborative software code creation (e.g. version control systems used by open source projects)

Putting these five factors together, one has a situation where a distributed, heterogenous group of passionate experts can work together, implementing collections of advanced algorithms on powerful computers, within software architectures modeled on human cognitive science. This is, I strongly suspect, how AGI is going to get created.

There has been quite amazing progress in all in the last two decades in all five of these areas. There's no comparison between the situation now, and the situation back in the 1960s and 1970s, or the late 1950s when the AI field was founded.

My first AI programs were written around 1980, on an Atari 400 computer with 8KB of RAM and a single 1-2Mhz processor. Presently, I'm running AI software on a machine with 8 3GHz processors and 96GB of RAM; plus an Nvidia GPU machine with 4 graphics cards, each having several hundred processors operating in parallel. These are not super-expensive supercomputers, they're both a couple years old and had cost in the ten thousand dollar range. My humble Macbook has 4GB of RAM and two 2GHz processors. These quantitative changes make a qualitative difference – the larger amounts of RAM available in current computers make it possible to create AI programs that maintain a lot of knowledge in a readily-available state, which is critical if one wants to try for advanced AGI. The faster processors available today make it possible to utilize this RAM effectively.

The process of programming is also totally different now than when I got started. Back then you had to write all your own code (except the programming language compiler and device drivers, although I had to write some drivers too now and then). Nowadays, most programming is a matter of hooking together bits and pieces that have each been written by a different person. Hard-core programmers frequently become frustrated that they don't need to write their own algorithms anymore, as best-of-breed algorithms are provided inside code libraries, with nice interfaces to allow them to be rapidly plugged into your own code. The implementations of the algorithms are regularly updated to keep pace with algorithmic and hardware improvements.

For example, recently, we have been modifying our OpenCog proto-AGI system to try to make it run better on machines with multiple processors. This has required a fair bit of work, but not as much as anticipated, because many of the data structures used are library functions that were already updated for efficient multiprocessor performance.

Back in the 80s, I did some amateur game programming, and it involved writing code that drew colors on individual pixels on the screen. Early AI programs involved computer-simulated "blocks worlds," which were extremely simplistic and allowed very few kinds of blocks manipulation. Now, in our OpenCog project, we are still using a sort of "blocks world" – but it's implemented in the Unity3D game engine, using an open-source Unity plug-in that makes Unity behave somewhat like the popular blocks-building game Minecraft.

In 1996, the first robot I built looked like an overturned salad bowl on three wheels. It had a bump sensor, a sonar rangefinder, and was a bit like an early-version Roomba that didn't vacuum. Presently, in OpenCog, we're working with Nao robots[1].

Nao Robots are short plastic humanoids that walk around and talk and actually "see" through their camera eyes. We'll soon be working with Hanson Robokind as well, who creates similar

---

[1] http://www.youtube.com/watch?v=2STTNYNF4lk

humanoids with various technical advantages over the current Nao bots, including emotionally expressive faces enabled by innovative artificial skin. Both Nao and Robokind robots are currently priced around $15,000. If you have a bigger budget, you can buy the PR2 for a mid-six-figure price tag, which lacks the humanoid appearance, but is very nearly capable of serving as a wheeled home service robot. All these robots can be simulated fairly effectively in freely available robot simulation software.

When I was in college in the early 80s, and poked around in the psychology section in the library, I found was an odd mixture of the "nebulous and unscientific" with the "extremely dry, boring and narrow-focused." Granted, there were the psychotherapists and other clinical psychologists, who were thinking about the whole mind, but in a quite impressionistic way that seemed more related to art or literature than science.

Freud and Jung and Maslow were fantastic to read, but it was impossible to take them seriously as guides for building AI. Clearly, they were just presenting an individual view on mind and life, much in the manner of philosophers like Plato or Nietzsche, or the medieval Buddhist psychologists. Sure, they were developing therapeutic methods that some people found helpful, but multiple, mutually contradictory religions have done that too. And "ability to help" is not a measure of underlying theoretical accuracy.

On the other hand, laboratory psychology experiments (whether on rats or humans) seemed to focus on very tiny issues like word priming or visual illusions, shying away from all the really interesting parts of the mind. As I later understood when I worked in a psychology department for a while (at the University of Western Australia, in the 1990s), this narrowness of focus was partly due to taste and partly due to practicality, as designing rigorous psychology lab experiments is really hard. Advances in neuroimaging have since helped experimental psychology a bit, but not as broadly as one might think, because of the lack of any non-invasive brain imaging technology that can measure the fine-grained temporal dynamics of numerous specific regions of a brain simultaneously.

I was frustrated that nobody seemed to be trying to put all the pieces together to try to really understand how the human mind worked, in a manner both holistic and scientific, using all the data available from all the different relevant disciplines: psychology, neuroscience, computer modeling, linguistics, philosophy, etc.

Well guess what? This happened; The interdisciplinary discipline of Cognitive Science emerged! In my academic days, I helped found two cognitive science degree programs: Waikato University in Hamilton, New Zealand, and the University of Western Australia in Perth.

Cognitive science certainly has not resolved all the questions about the mind, and, as a discipline, its success has been mixed. University cognitive science programs have all too often been absorbed into psychology departments, losing much of their cross-disciplinary flair and becoming little more than cognitive psychology. However, the cognitive science movement has made tremendous progress toward a unified understanding of the human mind.

And finally, all of these innovations – computing hardware, algorithms, cognitive science and robotics – have taken their current form in large part due to the Internet. Cloud computing, open-source software, downloadable code libraries, cross-disciplinary collaboration between scientists at different universities and in different nations, online virtual worlds and games – all this and more is enabled by Internet technology. It's not any one human nor any small team who's building the infrastructure to support AGI – it's the emerging Global Brain.

Putting together advances in computing hardware, robots and virtual worlds, algorithms and cognitive science, one arrives at a systematic and viable approach to advanced AGI. You start with a diagram explaining how the human mind works, and showing what are the main processes and how they Interact. Next, you look at the assemblage of existing algorithms and data structures, and figure out a set of algorithms and data structures that will do everything specified in the cognitive science diagram. Finally, you implement these algorithms and data structures in a way that can operate and is "able to scale" to run on modern networks of multiprocessor machines. Since this is a damn big job, you don't do it yourself, but you rely on a team of people communicating via the Internet, and drawing on expertise from others outside your team, in the form of Internet communications, online research papers and so forth.

This is what we're doing now with the OpenCog team. It's what a number of other research teams are doing too. According to my best understanding, it's how powerful AGI will finally be created.

Of course this isn't the only possible path. It's possible that detailed brain emulation will get there first. I'll focus on the integrative cognitive and computer science based approach here, because it's the one I'm following in most of my own work, and it's the one that I think has the greatest chance of rapid success. But ultimately, if either approach succeeds it will enable the other one. An AGI built via integrative cognitive and computer science will be a huge help in unraveling the mysteries of the brain; and an AGI built via emulating the brain would enable all sorts of experimentation that can't be done on biological brains, thus leading us to various less brain-like AGI architectures incorporating various computer science advances.

Naturally, each of the areas whose advance I've cited above also has its limitations. I think we're just barely at the point now where these various supporting disciplines have advanced enough to permit advanced AGI work. 10 years ago, creating advanced AGI might have been possible, but it would have been much more difficult to implement. 20 years ago it would have been a miracle. 10 years from now it will probably be a lot easier, due to better tools embodying richer understanding. 20 or 30 years from now, it may be a high school class project.

Computers are fast and RAM is abundant, but programming software for multiprocessor and distributed systems is still a pain. In a decade, this will be tremendously easier, due to more advanced software libraries and algorithms.

Algorithm and data structure libraries are great these days, but where AI-specific algorithms and structures are concerned, one often has to roll one's own. MATLAB is good for prototyping neural networks, but to build a scalable neural net with a novel architecture, you have to do your own coding. There are logic rule engines you can download and use to prototype inference ideas, but none of them scales up well, and all have fairly rigid restrictions. Whatever AI paradigms you prefer, chances are the available code libraries are not nearly up to the level of the best libraries for "standard computer science" algorithms and data structures. This is going to change, and it will eventually make implementation of fairly complex AI systems at least as simple as implementing database-back-ended websites is today.

It's amazing that we can experiment with humanoid robots now in a university research lab without massive funding – but these robots, cool as they are, have a lot of limitations. They can't walk outdoors without falling (Big Dog and some other robots can, but they're not humanoid and don't have good hands for manipulating things, and have other limitations). Their hands don't work that well. The Nao's camera eyes are not so great, at time of writing, though this may be remedied by the time you read these words.

And the video game and virtual worlds at our disposal today lack the richness of the real world – by a long shot. No fabrics, no richly dynamic fluids, no crap, no spit, no peanut butter – no oceans! When a game character picks up an object, this involves a preprogrammed invisible "socket" on the character's hand interacting with a preprogrammed invisible "socket" on the object

– it's not a free and flexible interaction like when a person picks up an object with their hand, or when a dog picks up a stick with its mouth. Robot simulation software doesn't have this limitation – but it's slow and not very scalable. Nobody has yet built a massive multiplayer robot simulator. But one thing is sure – they will.

Cognitive science, as exciting as it is and as fast as it has progressed, is still a lot stronger on structures than dynamics. In 1991 I joined a fantastically fun group of researchers called the "Society for Chaos Theory in Psychology", devoted to exploring implications of nonlinear dynamics for intelligence. The group still exists, and a portion of it, centered around Fred Abraham's lively "Winter Chaos Conference" in Vermont, maintains the revolutionary spirit the group had in the early 90s. One of the core ideas of this group is that most of the important features of intelligence are not static structural things, but rather dynamic attractors – complex emergent patterns of system dynamics that are neither stable nor repetitive nor random, but display more complex temporal structures. More and more data supporting this view has emerged, yet the mainstream of cognitive science hasn't yet incorporated these ideas into its thinking. The boxes and links in Figure 1 above aren't discrete brain regions with bundles of wires between them – in large part, they're dynamically assembled, continually self-organizing networks of neurons that maintain themselves via complex nonlinear "chaotic" dynamics.

Much of the reason cognitive science hasn't yet embraced nonlinear dynamics is that it's really hard to measure. We don't have good enough neuroimaging tools to measure the formation and interaction of chaotic attractors in the brain, and dynamical phenomena are pretty hard to study in psychology lab experiments. But as neuroimaging improves, our understanding of dynamics in the brain will improve, and cognitive science will get richer and deeper.

The Internet, as it stands, is nowhere near as great as it could be at fostering deep intellectual, scientific and engineering collaboration. It's hard to find the signal among the noise sometimes, and even with cheap multi-person videoconferencing like we have now, face-to-face meetings still have higher semantic and emotional bandwidth. Francis Heylighen at the Free University of Brussels – We'll talk to him later - is developing a host of new technologies aimed at accelerating the transformation of the Internet into a more richly intelligent Global Brain. This seems to be happening quite rapidly and effectively, and AGI research will benefit from this Global Brain, along with so many other pursuits.

As all these allied areas advance, AGI research will get easier and easier. So, if I wanted to make my job easier, I would just wait for the technology infrastructure to mature, and then start working on AGI some number of years from now. But of course, if I did that, somebody else might get there first!

Why do I care if someone else gets there first? Of course I have a fair dose of pride and ambition. It would be fun to be part of the first team to build AGI! Just as Edmund Hilary wanted to be the first to ascend Everest, although he also got a lot of pleasure just from the pure challenge of the climb itself. However, I have concerns that if AGI is developed too *late*, the risk of a dangerous outcome for humanity is greater.

If we can develop advanced AGI *soon*, then the chance of a young AGI somehow spiraling out of human control, or being rapidly deployed by evil humans for massive destruction, seems fairly low. For a new AGI to be used in destructive ways anytime soon (now, or in the near future), it would require the use of a lot of complex, slow-moving infrastructure, and require the participation and coordination of a lot of people. On the other hand, once there is a lot more advanced technology of various sorts available, it could well be possible for a young AGI to wreak a lot of damage very quickly. For these reasons, I think it's a better idea to bring our first baby non-human AGI into the world when the "toys" it will be interacting with are a bit less potent. It would be a little like giving a toddler a loaded gun, but without the risk of the toddler shooting itself.

Why So Little Work on AGI?

So now you're thinking "Well, if all that's all true, and we're poised for an AGI revolution, then how come AGI isn't the focus of a trillion dollar industry right now? How come AGI isn't the largest, best-funded department at MIT and Caltech?

However, if you're old enough to have been there, or are familiar enough with the history of the internet to be able to imagine – Let's go back to 1990. How much thought and research was going into the Internet and the Web? Hardly any – compared to all the energy and resources on the planet at the time. Yet the world was poised for it – and obviously so, in hindsight. The world has a way of missing the obvious. But the cool thing is that when the obvious rears its head, wakes up the world, and bites it on its ass, the world tends to come around very quickly.

For these reasons, the people of the world *will be more than ready for AGI, when it finally happens*. All the sci fi movies and television shows of the past 100 years, and the exponential rate of scientific advancement will have prepared them adequately to put it to good use.

In the case of AGI, there are various historical and practical factors encouraging most of the world to miss the obvious. Peter Voss, an AI theorist, entrepreneur and futurist, has summarized the situation well. His observation, back in 2002, was that, of all the scientists and engineers working in the AI field,

1. 80% don't believe in the concept of General Intelligence (but instead, in a large collection of specific skills & knowledge).

2. Of those that do, 80% don't believe it's possible—either ever, or for a long, long time.
3. Of those that do, 80% work on domain-specific AI projects for reasons of commercial or academic politics (results are a lot quicker).
4. Of those left, 80% have the wrong conceptual framework.
5. And nearly all of the people operating under basically correct conceptual premises lack the resources to adequately realize their ideas.

I think Peter's argument is basically on-target. Of course, the 80% numbers are crude approximations, and most of the concepts involved are fuzzy in various ways. But an interesting observation is that, whatever the percentages actually are, most of them have decreased considerably since 2002. Today, compared to in 2002, many more AI researchers believe AGI is a feasible goal to work towards, and that it might arise within their lifetimes. The researchers participating in the AGI conference series have mostly bypassed the first 2 items on Peter's list. And while funding for AGI research is still very difficult compared to some other research areas, the situation has definitely improved in the last 10 years.

From my personal point of view as an AGI researchers, the most troubling of Peter's five points is the fifth one. Most AI researchers who believe AGI is feasible – and would like to be spending their lives working on it – are still working on highly domain-specific AI projects with much of their time, because that's where the funding is. Even scientists need to eat, and AGI research requires computers and programmers and so forth. I'm among the world's biggest AGI advocates, and I myself spend about half my time on AGI research and the other half on narrow – AI projects that bring me revenue to pay my mortgage and put my kids through college.

Narrow AI gets plenty of money, in forms like Google's and Microsoft's expenditure on AI-based Web search and ad placement, and the military's expenditure on AI-based intelligence analysis and unmanned vehicle control. But AGI is relatively minimally funded, compared to these shorter-term, narrower-scope AI applications.

Indeed, from a practical business perspective, at this stage AGI is research with at best a medium term payoff – it's not going to make anyone's profits higher next quarter. One can potentially chart paths that transition from Narrow AI to AGI, and this may be a viable way to get to advanced AGI, but it's certainly not the fastest or easiest way – and it's different than what would happen if society were to explicitly fund AGI research in a big way.

The relatively paltry funding of AGI isn't just due to its speculative nature – society is currently willing to fund a variety of speculative science and engineering projects. Billion dollar particle accelerators, space exploration, the sequencing of human and animal genomes, stem cell research, and so forth. If these sorts of projects merit Big Science level funding, why is AGI research left out? After all, the potential benefits are obviously tremendous. There are potential dangers too, to be sure – but there are also clear potential dangers of particle physics research (discovering better bombs is arguably a hazardous pursuit), and that doesn't stop us.

Any social phenomenon has multiple intertwined causes, but the main reason for the AGI field's relatively paltry funding stream is probably negative momentum from the failures of the original generation of AI researchers. The AI gurus of the 1960s were claiming they could create human-level AI within a decade or less. They were wrong – they lacked the needed hardware, their software tools were primitive, and their conceptual understanding of intelligence was too crude. Just because they were wrong then, doesn't mean the current AGI field is similarly wrong – but the "guilt by association" lingers.

There is an interesting analogy here, related to my observation above that almost nobody was hyping or funding the Web in 1990. Think about the early visionaries who foresaw the Web – Vannevar Bush in the 1950s, Ted Nelson in the 1960s, and others. They understood the potential computer technology held to give rise to something like today's Web – and Ted Nelson even tried to get something Web-like built, back before 1970. But the technology just wasn't there to support his vision. In principle it might have been do-able doing the technology of that era, but it would have been insanely difficult – whereas by the time the Web came about in the 1990s, it seemed almost a natural consequence of the technological infrastructure existing at that time. Similarly, in the 1960s, even if someone had come up with a workable design for human-level AGI, it would have been extraordinarily difficult to get it implemented and working using the hardware and software tools available. But now, with cloud computing, multiprocessor machines with terabytes of RAM, powerful algorithm libraries and debuggers, and a far more mature theory of cognitive science, we are in a whole different position. The conceptual and technological ecosystem is poised for AGI, in the same sense that it was poised for the Web in the 1990s. And just as the Web spread faster than almost anybody foresaw – so will AGI, once it gets started.

Shifting to a different historical analogy, I think about the future of AGI as falling into two phases – before and after the "AGI Sputnik" event.

When the Russians launched Sputnik, this sent a message to the world: "Wow! Going into space is not only a possibility; it's a dramatic reality! The time for humanity to explore space is now!" The consequence was the space race, and the rise of modern space technology.

Similarly, at some point, some AGI research team is going to produce a computer program or robot that does something that makes the world wake up and say: "Wow! Genuinely smart AI is not just a possibility; it's a dramatic reality! The time for humanity to create smart machines is now!" At that point, government and industry will put themselves fully behind the creation of advanced AGI – and progress will accelerate tremendously. The potential of AGI to benefit humanity, and every nation and corporation, is evidently humongous – and all that's needed to shift things dramatically from the current funding regime to something overwhelmingly different, is one single crystal-clear demonstration that human-level AGI is feasible in the near term. We don't have that demonstration right now, but I – and a number of other AGI researchers – believe we know how to do it... And I'm betting it will happen within the lifetimes of most people reading this book.

In the late 60s and early 70s, in the era of the Apollo moon missions, I and every other little American kid wanted to grow up and be an astronaut. After the AGI Sputnik event, kids will want to grow up and be AGI developers – or AGIs!

I believe my OpenCog AGI approach has what it takes to get us to an AGI Sputnik event – in the form, perhaps, of a video game character or a humanoid robot that holds meaningful conversations about the things in its environment. Imagine talking to a robot that really gives you sense that it knows what it's talking about – that it understand what it's doing, and knows who it is and who you are. That will be a damn strange feeling – and a wonderful one. And everyone who gets that feeling will understand that humanity is about to take the next, huge, step.

I'll talk a bit about OpenCog in these pages – but my main point here isn't to sell my own particular approach to AGI; rather, to talk about AGI in general, and the broader implications that the advent of AGI will have for humanity and beyond. Many of my colleagues have their own different perspective on the optimal technical approach to create AGI. My main goal here is, first, to get across the points that AGI is probably coming fairly soon, and it's going to be a huge change and quite possibly a fantastic one for all of us; and second, to explore some of the things that AGI has to teach us about the nature of mind and intelligence.

AGI is coming – probably faster than you think – and it's going to be ***really, really interesting…***

4

Ten Years to the Singularity If We Really Really Try

We've discussed the Vinge-ean, Kurzweil-ian argument that human-level AGI may be upon us shortly. By extrapolating various key technology trends into the near future, in the context of the overall dramatic technological growth the human race has seen in the past centuries and millennia, it seems quite plausible that superintelligent artificial minds will be here much faster than most people think.

This sort of objective, extrapolative view of the future has its strengths, and is well worth pursuing. But I think it's also valuable to take a more subjective and psychological view, and think about AGI and the Singularity in terms of the power of the human spirit; What we really want for ourselves, and what we can achieve if we really put our minds to it.

I presented this sort of perspective on the timeline to Singularity and advanced AGI at the TransVision 2007 futurist conference, in a talk called "Ten Years to a Positive Singularity (If We Really, Really Try)." The conference was in Helsinki, Finland, and I wasn't able to attend in person so I delivered the talk by video – if you're curious you can find it online[2].

The basic point of the talk was that if society put the kind of money and effort into creating a positive Singularity that we put into things like wars or television shows, then some pretty amazing things might happen. To quote a revised version of the talk, given to a different audience just after the financial crisis of Fall 2008:

---

[2] http://www.youtube.com/watch?v=BelOkx4Jxyg

Look at the US government's response to the recent financial crisis – suddenly they're able to materialize a trillion dollars here, a trillion dollars there. What if those trillions of dollars were being spent on AI, robotics, life extension, nanotechnology and quantum computing? It sounds outlandish in the context of how things are done now – but it's totally plausible.

If we made a positive Singularity a real focus of our society, I think a ten year time-frame or less would be eminently possible.

Ten years from now would be 2020. Ten years from 2007, when that talk was originally given, would have been 2017, only 7 years from now. Either of these is a long time before Kurzweil's putative 2045 prediction. Whence the gap?

When he cites 2045, Kurzweil is making a guess of the "most likely date" for the Singularity. The "ten more years" prediction is a guess of how fast things could happen with an amply-funded, concerted effort toward a beneficial Singularity. So the two predictions have different intentions.

We consider Kurzweil's 2045 as a reasonable extrapolation of current trends, but we also think the Singularity could come a lot sooner, or a lot later, than that.

How could it come a lot later? Some extreme possibilities are easy to foresee. What if terrorists nuke the major cities of world? What if anti-technology religious fanatics take over the world's governments? But less extreme outcomes could also occur, with similar outcomes. Human history could just take a different direction than massive technological advance, and be focused on warfare, or religion, or something else.

Or, though we reckon this less likely, it is also possible we could hit up against tough scientific obstacles that we can't foresee right now. Intelligence could prove more difficult for the human brain to puzzle out, whether via analyzing neuroscience data, or engineering intelligent systems.

Moore's Law and its cousins could slow down due to physical barriers, designing software for multicore architectures could prove problematically difficult; the pace of improvement in brain scanners could slow down.

How, on the other hand, could it take a lot less time? If the right people focus their attention on the right things.

The Ten Years to the Singularity talk began with a well-known motivational story, about a guy named George Dantzig (no relation to the heavy metal singer Glenn Danzig!). Back in 1939, Dantzig was studying for his PhD in statistics at the University of California, Berkeley. He arrived late for class one day and found two problems written on the board. He thought they were the homework assignment, so he wrote them down, then went home and solved them. He thought they were particularly hard, and it took him a while. But he solved them, and delivered the solutions to the teacher's office the next day. Turns out, the teacher had put those problems on the board as examples of "unsolvable" statistics problems; Two of the greatest unsolved problems of mathematical statistics in the world, in fact. Six weeks later, Dantzig's professor told him that he'd prepared one of his two "homework" proofs for publication. Eventually, Dantzig would use his solutions to those problems for his PhD thesis.

Here's what Dantzig said about the situation: "If I had known that the problems were not homework, but were in fact two famous unsolved problems in statistics, I probably would not have thought positively, would have become discouraged, and would never have solved them."

Dantzig solved these problems because he thought they were solvable; He thought that other people had already solved them.

He was just doing them as "homework," thinking everyone else in his class was going to solve them too.

There's a lot of power in expecting to win. Athletic coaches know about the power of streaks. If a team is on a roll, they go into each game expecting to win, and their confidence helps them see more opportunities to win. Small mistakes are just shrugged away by the confident team, but if a team is on a losing streak, they go into each game expecting to screw up, somehow. A single mistake can put them in a bad mood for the whole game, and one mistake can pile on top of another more easily.

To take another example, let's look at the Manhattan Project. America thought they needed to create nuclear weapons before the Germans did. They assumed it was possible, and felt a huge burning pressure to get there first. Unfortunately, what they were working on so hard, with so much brilliance, was an ingenious method for killing a lot of people. But, whatever you think of the outcome, there's no doubt the pace of innovation in science and technology in that project was incredible, and it all might have never happened if the scientists involved didn't already believe that Germany was ahead of them, and that somehow their inventing the ability to kill thousands, first, would save humanity.

How Might a Positive Singularity Get Launched In 10 Years From Now?

This way of thinking leads to a somewhat different way of thinking about the timing of the Singularity. What if, rather than thinking about it as a predictive exercise (an exercise in objective studying what's going to happen in the world, as if we were outsiders to the world). What if we thought about it the way an athlete thought about a game when going into it, or the way the Manhattan Project scientists thought at the start of the project, or the way Dantzig thought about his difficult homework problems?

What if we knew it was possible to create a positive Singularity in ten years? What if we assumed we were going to win, as a provisional but reasonable hypothesis?

What if we thought everyone else in the class knew how to do it already?

What if we were worried the bad guys were going to get there first?

Under this assumption, how then would we go about trying to create a positive Singularity?

Following this train of thought, even just a little ways, will lead you along the chain of reasoning that led us to write this book.

One conclusion that seems fairly evident when taking this perspective is that AI is the natural area of focus.

Look at the futurist technologies at play these days:

- nanotechnology
- biotechnology
- robotics
- AI

and ask, “which ones have the most likelihood of bringing us a positive Singularity within the next ten years?”

Nano and bio and robotics are all advancing fast, but they all require a lot of hard engineering work.

AI requires a lot of hard work too, but it’s a softer kind of hard work. Creating AI relies only on human intelligence, not on painstaking and time-consuming experimentation with physical substances and biological organisms.

And how can we get to AI? There are two big possibilities:

- Copy the human brain, or
- Come up with something cleverer

Both approaches seem viable, but the first approach has a problem. Copying the human brain requires far more understanding of the brain than we have now. Will biologists get there in ten years from now. Probably not. Definitely not in five years.

So we're left with the other choice, come up with something cleverer. Figure out how to make a thinking machine, using all the sources of knowledge at our disposal: Computer science and cognitive science and philosophy of mind and mathematics and cognitive neuroscience and so forth.

But if this is feasible to do in the near term, which is what we're suggesting, then why don't have AI's smarter than people right now? Of course, it's a lot of work to make a thinking machine, but making cars and rockets and televisions is also a lot of work, and society has managed to deal with those problems.

The main reason we don't have real AI right now is that almost no one has seriously worked on the problem. And (here is where things get even more controversial!) most of the people that have worked on the problem have thought about it in the wrong way.

Some people have thought about AI in terms of copying the brain, but, as I mentioned earlier, that means you have to wait until the neuroscientists have finished figuring out the brain. Trying to make AI based on our current, badly limited understanding of the brain is a clear recipe for failure. We have no understanding yet of how the brain represents or manipulates abstraction. Neural network AI is fun to play with, but it's hardly surprising it hasn't led to human-level AI yet. Neural nets are based on extrapolating a very limited understanding of a few very narrow aspects of brain function.

The AI scientists who haven't thought about copying the brain, have mostly made another mistake; They've thought like computer scientists. Computer science is like mathematics; It's all about elegance and simplicity. You want to find beautiful, formal solutions. You want to find a single, elegant principle. A single structure. A single mechanism that explains a whole lot of different things. A lot of modern theoretical physics is in this vein. The physicists are looking for a single, unifying equation underlying every force in the universe. Well, most computer scientists working on AI are looking for a single algorithm or data structure underlying every aspect of intelligence.

But that's not the way minds work. The elegance of mathematics is misleading. The human mind is a mess, and not just because evolution creates messy stuff. The human mind is a mess because intelligence, when it has to cope with limited computing resources, is necessarily messy and heterogenous.

Intelligence does include a powerful, elegant, general problem-solving component, and some people have more of it than others. Some people I meet seem to have almost none of it at all.

But intelligence also includes a whole bunch of specialized problem-solving components dealing with things like: vision, socialization, learning physical actions, recognizing patterns in events over time, and so forth. This kind of specialization is necessary if you're trying to achieve intelligence with limited computational resources.

Marvin Minsky has introduced the metaphor of a society. He says a mind needs to be a kind of society, with different agents carrying out different kinds of intelligent actions and all interacting with each other.

But a mind isn't really like a society. It needs to be more tightly integrated than that. All the different parts of the mind, parts which are specialized for recognizing and creating different kinds of patterns, need to operate very tightly together, communicating in a common language, sharing information, and synchronizing their activities.

And then comes the most critical part; The whole thing needs to turn inwards on itself. Reflection. Introspection. These are two of the most critical kinds of specialized intelligence that we have in the human brain, and both rely critically on our general intelligence ability. A mind, if it wants to be really intelligent, has to be able to recognize patterns in itself, just like it recognizes patterns in the world, and it has to be able to modify and improve itself based on what it sees in itself. This is what "self" is all about.

This relates to what the philosopher Thomas Metzinger calls the "phenomenal self ." All humans carry around inside our minds a "phenomenal self." An illusion of a holistic being. A whole person. An internal "self" that somehow emerges from the mess of information and dynamics inside our brains. This illusion is critical to what we are. The process of constructing this illusion is essential to the dynamics of intelligence.

Brain theorists haven't understood the way the self emerges from the brain yet, because brain mapping isn't advanced enough.

Computer scientists haven't understood the self, because it isn't about computer science. It's about the emergent dynamics that happen when you put a whole bunch of general and specialized pattern recognition agents together; A bunch of agents created in a way that they can really cooperate, and when you include in the mix agents oriented toward recognizing patterns in the society as a whole.

The specific algorithms and representations inside the pattern recognition agents – algorithms dealing with reasoning, or seeing, or learning actions, or whatever – these algorithms are what computer science focuses on. They're important, but they're not really the essence of intelligence. The essence of intelligence lies in getting the parts to all work together in a way that gives rise to the phenomenal self. That is, it's about wiring together a collection of structures and processes into a holistic system in a way that lets the whole system recognize significant patterns in itself. With very rare exceptions, this has simply not been the focus of AI researchers.

When talking about AI in these pages, I'll use the word "patterns" a lot. This is inspired in part by my book "The Hidden Pattern," which tries to get across the viewpoint that everything in the universe is made of patterns. This is not a terribly controversial perspective – Kurzweil has also described himself as a "patternist." In the patternist perspective, everything you see around you, everything you think, everything you remember; That's a pattern!

Following a long line of other thinkers in psychology and computer science, we conceive intelligence as the ability to achieve complex goals in complex environments. Even complexity itself has to do with patterns. Something is "complex" if it has a lot of patterns in it.

A "mind" is a collection of patterns for effectively recognizing patterns. Most importantly, a mind needs to recognize patterns about what actions are most likely to achieve its goals.

The phenomenal self is a big pattern, and what makes a mind really intelligent is its ability to continually recognize this pattern; The phenomenal self in itself.

Does it Take a Manhattan Project?

One of the more interesting findings from the "How Long Till Human-Level AI" survey we discussed above was about funding, and the likely best uses of hypothetical massive funding to promote AGI progress.

In the survey, we used the Manhattan Project as one of our analogies, just as I did in part of the discussion above -- but in fact, it may be that we don't need a Manhattan Project scale effort to get Singularity-enabling AI. The bulk of AGI researchers surveyed at AGI-09 felt that, rather than a single monolithic project, the best use of massive funding to promote AGI would be to fund a heterogenous pool of different projects, working in different but overlapping directions.

In part, this reflects the reality that most of the respondents to the survey probably thought they had an inkling (or a detailed understanding) of a viable path to AGI, and feared that an AGI Manhattan Project would proceed down the wrong path instead of their "correct" path. But it also reflects the realities of software development. Most breakthrough software has come about through a small group of very brilliant people working together very tightly and informally. Large teams work better for hardware engineering than software engineering.

It seems most likely that the core breakthrough enabling AGI will come from a single, highly dedicated AGI software team. After this breakthrough is done, a large group of software and hardware engineers will probably be useful for taking the next step, but that's a different story.

What this suggests is that, quite possibly, all we need right now to get Singularity-enabling AGI is to get funding to a dozen or so of the right people. This would enable them to work on the right AGI project full time for a decade or so, or maybe even less.

It's worth emphasizing that my general argument for the potential imminence of AGI does not depend on my perspective on any particular route to AGI being feasible. Unsurprisingly, I'm a big fan of the OpenCog project, of which I'm one of the founders and leaders. I'll tell you more about this a little later on. But you don't need to buy my argument for OpenCog as the most likely path to AGI, in order to agree with my argument for creating a positive Singularity by funding a constellation of dedicated AGI teams.

Even if OpenCog were the wrong path, there could still be a lot of sense in a broader bet that funding 100 dedicated AGI teams to work on their own independent ideas will result in one of them making the breakthrough. What's shocking, given the amount of money and energy going into other sorts of technology development, is that this isn't happening right now. (Or maybe it is, by the time you are reading this!!)

Keeping it Positive

I've talked more about AI than about the Singularity or positiveness. Let me get back to those.

It should be obvious that if you can create an AI vastly smarter than humans, then pretty much anything is possible.

Or at least, once we reach that stage, there's no way for us, with our puny human brains, to really predict what is or isn't possible. Once the AI has its own self, and has superhuman level intelligence, it's going to start learning and figuring things out on its own.

But what about the "positive" part? How do we know this AI won't annihilate us all? Why won't it just decide we're a bad use of mass-energy, and re-purpose our component particles for something more important?

There's no guarantee of this not happening, of course.

Just like there's no guarantee that some terrorist won't nuke your house tonight, or that you won't wake up tomorrow morning to find the whole life you think you've experienced has been a long strange dream. Guarantees and real life don't match up very well. (Sorry to break the news.)

However, there are ways to make bad outcomes unlikely, based on a rational analysis of AI technology and the human context in which it's being developed.

The goal systems of humans are pretty unpredictable, but a software mind can potentially be different, because the goal system of an AI can be more clearly and consistently defined. Humans have all sort of mixed-up goals, but there seems no clear reason why one can't create an AI with a more crisply defined goal of helping humans and other sentient beings, as well as being good to itself. We will return to the problem of defining this sort of goal more precisely later, but to make a long story short, one approach (among many) is to set the AI the goal of figuring out what is most common among the various requests that various humans may make of it.

One risk of course is that, after it grows up a bit, the AGI changes its goals, even though you programmed it not to. Every programmer knows you can't always predict the outcome of your own code. But there are plenty of preliminary experiments we can do to understand the likelihood of this happening. And there are specific AGI designs, such as the GOLEM design we'll discuss below, that have been architected with a specific view toward avoiding this kind of pathology. This is a matter best addressed by a combination of experimental science and mathematical theory, rather than armchair philosophical speculation.

Ten Years to the Singularity?

How long till human-level or superhuman AGI? How long till the Singularity? None of us knows. Ray Kurzweil, and others, have made some valuable projections and predictions. But you and I are not standing outside of history analyzing its progress; we are co-creating it. Ultimately the answer to this question is highly uncertain, and, among many other factors, it depends on what **we** do.

To quote the closing words of the Ten Years to The Singularity TransVision talk:

> *A positive Singularity in 10 years?*

Am I sure it's possible? Of course not.

But I do think it's plausible.

And I know this: If we assume it isn't possible, it won't be.

And if we assume it is possible – and act intelligently on this basis – it really might be. That's the message I want to get across to you today.

There may be many ways to create a positive Singularity in ten years. The way I've described to you – the AI route – is the one that seems clearest to me. There are six billion people in the world so there's certainly room to try out many paths in parallel.

But unfortunately the human race isn't paying much attention to this sort of thing. Incredibly little effort and incredibly little funding goes into pushing toward a positive Singularity. I'm sure the total global budget for Singularity-focused research is less than the budget for chocolate candy – let alone beer ... Or TV... Or weapons systems!

I find the prospect of a positive Singularity incredibly exciting – and I find it even more exciting that it really, possibly could come about in the next ten years. But it's only going to happen quickly if enough of the right people take the right attitude – and assume it's possible, and push for it as hard as they can.

Remember the story of Dantzig and the unsolved problems of statistics. Maybe the Singularity is like that. Maybe superhuman AI is like that. If we don't think about these problems as impossibly hard – quite possibly they'll turn out to be solvable, even by mere stupid humans like us.

This is the attitude I've taken with my work on OpenCog. It's the attitude Aubrey de Grey has taken with his work on life extension. The more people adopt this sort of attitude, the faster the progress we'll make.

We humans are funny creatures. We've developed all this science and technology – but basically we're still funny little monkeylike creatures from the African savannah. We're obsessed with fighting and reproduction and eating and various apelike things. But if we really try, we can create amazing things – new minds, new embodiments, new universes, and new things we can't even imagine.

5

How Long Till AGI?

with Ted Goertzel and Seth Baum

This chapter originated as an H+ Magazine article. A longer, more technical version appeared in the journal "Technological Forecasting and Social Change".

What do other experts think about Kurzweil's projection of an AI-powered Singularity around 2045?

Opinions are all over the map – but one interesting data point is a survey that Seth Baum, my father Ted Goertzel and I conducted at the AGI-09 conference, a gathering of Artificial General Intelligence researchers in Washington DC in March 2009, on the specific topic of "How Long Till Human Level AI?"

Rather than a shallow survey of a large number of people, this was an in-depth survey of a small number of experts – what is known in the trade as an "expert elicitation." The sample of experts was avowedly biased – most (though not all) researchers who would bother to attend an AGI conference are relatively optimistic about the near-term feasibility of AGI. But even given this bias, it's very interesting to sample the perspective of AGI-interested experts and see where there opinions fall on various aspects of the question "when will human-level AI be here?"

Two Earlier Surveys of Expert Opinion on the Timing of Human-Level AI

We know of two previous studies exploring expert opinion on the future of artificial general intelligence. In 2006, a seven-question poll was taken of participants at the AI@50 conference. Four of the seven questions are particularly relevant. Asked "when will computers be able to simulate every aspect of human intelligence?", 41% said "More than 50 years" and 41% said "Never". Seventy-nine percent said an accurate model of thinking is impossible without further discoveries about brain functioning. Thus approximately 80% of the participants at the conference were AI "timing pessimists" and 20% "timing optimists," percentages that may reflect the wider pattern in the field. Sixty percent of the participants strongly agreed that "AI should take a multidisciplinary approach, incorporating stats, machine learning, linguistics, computer science, cognitive psychology, philosophy, and biology". And finally, 71% said that "statistical/probabilistic methods are most accurate in representing how the brain works".

The other survey, taken in 2007 by futurist entrepreneur Bruce Klein, was an online survey that garnered 888 responses, asking one question: "When will AI surpass human-level intelligence?" Most of those who chose to respond to this survey were AI "optimists" who believed that human-level artificial intelligence would be achieved during the next half century. The distribution of responses is shown below:

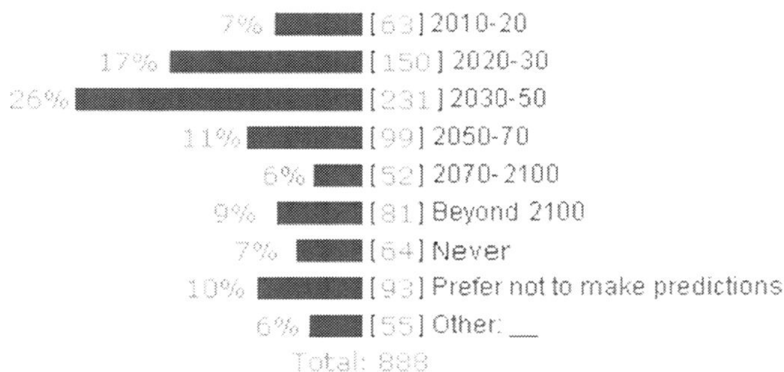

The AI@50 and Klein studies are interesting because they show that significant numbers of experts and interested persons believe that AGI with intelligence at least equaling the human level will exist within upcoming decades. Our own study probes more deeply into the thinking of people with substantial AGI expertise, and comes up with similar findings, but giving more details about the experts' underlying beliefs.

The Questions Asked of the Experts

The first set of questions in our survey elicited experts' beliefs about when AI would reach each of four milestones: passing the Turing test,[3] performing Nobel quality work, passing third grade, and becoming superhuman. These specific milestones were selected to span a variety of levels of advanced general intelligence.

For each milestone, two question versions were asked – with and without massive additional funding – making for eight total milestone questions. These two versions explore the possibility that when the milestones are reached depends on the resources available to researchers. The amount of additional funding listed, $100 billion per year, is obviously more than could be used for AGI research; the intent with this figure was to ensure that money would not be a scarce resource in this hypothetical AGI development scenario.

(Statistics geeks may be interested that, for each of the eight milestone questions, we asked the respondents to give us estimates representing 10%, 25%, 75% and 90% confidence intervals, as well as their best estimate dates. In the full write-up of our report we will present this data, but in this article we'll stick with the "best estimates," for sake of simplicity.)

Our next set of questions covered four topics. Three questions asked what embodiment the first AGIs would have: physical or

---

[3] While taking the survey, several participants expressed concern that the Turing test milestone is ambiguous, due to the numerous variations of the test. In response to this, several participants specified the Turing test variant their responses are based on. At the time of survey distribution, a verbal suggestion was given to consider the "one hour" rather than the "five minute" Turing test as some potential participants felt the latter could too easily be "gamed" by narrow-AI chatbots without significant general intelligence.

virtual robot bodies or a minimal text- or voice-only embodiment. Eight questions asked what AI software paradigm the first AGIs would be based on: formal neural networks, probability theory, uncertain logic, evolutionary learning, a large hand-coded knowledge-base, mathematical theory, nonlinear dynamical systems, or an integrative design combining multiple paradigms. Three questions asked the likelihood of a strongly negative-to-humanity outcome if the first AGIs were created by: an open-source project, the US military, or a private for-profit software company. Two true/false questions asked if quantum computing or hypercomputing would be required for AGI. Two yes/no questions asked if AGIs emulating the human brain conceptually or near-exactly would be conscious in the sense that humans are. Finally, 14 questions asked experts to evaluate their expertise in: cognitive science, neural networks, probability theory, uncertain logic, expert systems, theories of ethics, evolutionary learning, quantum theory, quantum gravity theory, robotics, virtual worlds, software engineering, computer hardware design, and cognitive neuroscience.

We posed these questions to 21 participants, with a broad range of backgrounds and experience, all with significant prior thinking about AGI. Eleven are in academia, including six Ph.D. students, four faculty members and one visiting scholar, all in AI or allied fields. Three lead research at independent AI research organizations, and three do the same at information technology organizations. Two are researchers at major corporations. One holds a high-level administrative position at a relevant non-profit organization. One is a patent attorney. All but four participants reported being actively engaged in conducting AI research.

What the Experts Said About The Timing of Human-Level AI

Unsurprisingly, our results showed that the majority of the experts who participated in our study were AGI timing optimists – but there was a significant minority of timing pessimists. Although it's worth noting that *all* the experts in our study, even the most pessimistic ones, gave at least a 10% chance of some AI milestones being achieved within a few decades.

The range of best-guess time estimates for the AGI milestones without massive additional funding is summarized below:

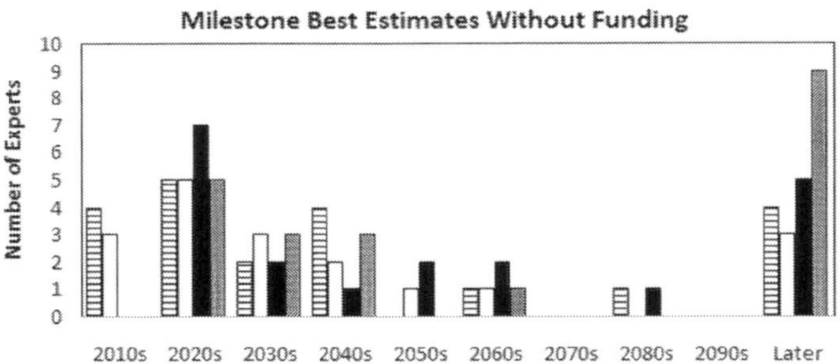

Distributions of best guess estimates for when AI would achieve four milestones without additional funding: the Turing Test (horizontal lines), third grade (white), Nobel-quality work (black), and superhuman capability (grey).

Another interesting result concerns the role of funding – the predicted impact on the AGI field of a hypothetical multibillion dollar funding infusion. Most experts estimated that a massive funding increase of this nature would cause the AI milestones to be reached sooner. However, for many of these experts, the difference in milestone timing with and without massive funding was quite small – just a few years. Furthermore, several experts estimated that massive funding would actually cause the AI milestones to be reached later. One reason given for this is that with so much funding, "many scholars would focus on making money and administration" instead of on research. Another is that "massive funding increases corruption in a field and its oppression of dissenting views in the long term". Of those who thought funding would make little difference, a common reason was that AGI progress requires theoretical breakthroughs from just a few dedicated, capable researchers, something that does not depend on massive funding. Another common reason was that the funding would not be wisely targeted. Several noted that funding could be distributed better if there was a better understanding of what paradigms could produce AGI, but such an understanding is either not presently available or not likely to be understood by those who would distribute funds.

Because of the lack of agreement on a single paradigm, several experts recommended that modest amounts of funding should be distributed to a variety of groups following different approaches, instead of large amounts of funding being given to a "Manhattan Project" type crash program following one approach. Several also observed that well-funded efforts guided by a single paradigm had failed in the past, including the Japanese Fifth Generation Computer Systems project. On this, one said, "AGI requires more theoretical study than real investment." Another said, "I believe the development of AGI's to be more of a tool and evolutionary problem than simply a funding problem. AGI's will be built upon tools that have been developed from previous tools. This evolution in tools will take time. Even with a crash project and massive funding, these tools will still need time to develop and mature." Given that these experts are precisely those who would benefit most from increased funding, their skeptical views of the impact of hypothetical massive funding are very likely sincere.

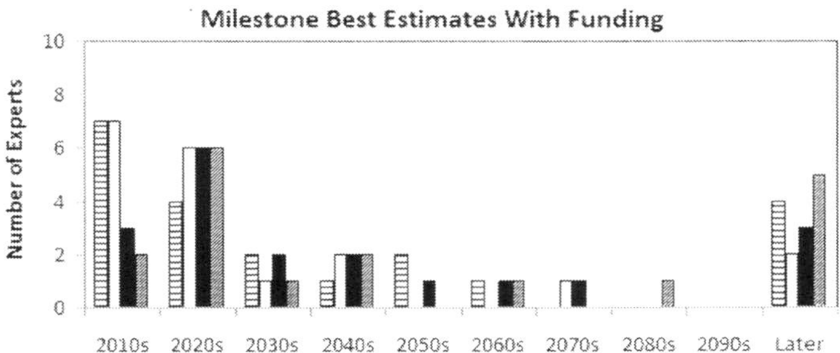

Distributions of best guess estimates for when AI would achieve four milestones with additional funding: the Turing Test (horizontal lines), third grade (white), Nobel-quality work (black), and superhuman capability (grey).

There was substantial agreement among the more optimistic participants in the study. Most thought there was a substantial likelihood that the various milestones would be passed sometime between 2020 and 2040. 16 experts gave a date before 2050 as their best estimate for when the Turing test would be passed. 13 experts gave a date between 2020 and 2060 as their best estimate for when superhuman AI would be achieved. (The others all estimated later dates, ranging from 2100 to never.) The opinions of the optimists were similar to and even slightly more optimistic than Kurzweil's well-known projections.

As noted above, optimism in this sense is about the timing of AI milestones. It does not imply a belief that achieving AGI would be a good thing. To the contrary, one can be optimistic that AGI will happen soon yet believe that AGI would have negative outcomes. Indeed, several experts reported this pair of beliefs. Results about the likelihood of negative outcomes are discussed further below.

What Kind of Human-Level AI Will Come First?

To our surprise, there was not much agreement among experts on the order in which the four milestones (Turing test; third grade; Nobel; superhuman) would be achieved. There was consensus that the superhuman milestone would be achieved either last or at the same time as other milestones. However, there was significant divergence regarding the order of the other three milestones. One expert argued that the Nobel milestone would be easier precisely because it is more sophisticated: to pass the Turing test, an AI must "skillfully hide such mental superiorities". Another argued that a Turing test-passing AI needs the same types of intelligence as a Nobel AI "but additionally needs to fake a lot of human idiosyncrasies (irrationality, imperfection, emotions)". Finally, one expert noted that the third grade AI might come first because passing a third grade exam might be achieved "by advances in natural language processing, without actually creating an AI as intelligent as a third-grade child". This diversity of views on milestone order suggests a rich, multidimensional understanding of intelligence. It may be that a range of milestone orderings are possible, depending on how AI development proceeds.

The milestone order results highlight the fact that many experts do not consider it likely that the first human-level AGI systems will closely mimic human intelligence. Analogy to human intelligence would suggest that achieving an AGI capable of Nobel level science would take much longer than achieving an AGI capable of conducting a social conversation. However, as discussed above, an AGI would not necessarily mimic human intelligence. This could enable it to achieve the intelligence milestones in other orders.

What Kind of Technical Approach Will First Achieve Human-Level AI?

Our initial survey asked about eight technical approaches used in constructing AI systems: formal neural networks, probability theory, uncertain logic, evolutionary learning, a large hand-coded knowledge-base, mathematical theory, nonlinear dynamical systems, and integrative designs combining multiple paradigms. For each, it requested a point estimate of the odds that the approach would be critical in the creation of human-level AGI. They survey also asked about the odds that physical robotics, virtual agent control, or minimal text- or voice-based embodiments would play a critical role.

The question about robotics vs. other forms of embodiment received a range of responses. There were responses of .9, .94, .89 and .6 for physical robot embodiment, but the mean response was only 27%. The handful of participants who felt robotics was crucial were all relatively optimistic. The preliminary impression one obtains is that a few researchers are highly bullish on robotics as the correct path to AGI in the relatively near term, whereas the rest feel robotics is probably not necessary for AGI.

Impacts of AGI

In science fiction, intelligent computers frequently become dangerous competitors to humanity, sometimes even seeking to exterminate humanity as an inferior life form. And indeed, based on our current state of knowledge, it's hard to discount this as a real possibility, alongside much more benevolent potential outcomes. We asked experts to estimate the probability of a negative-to-humanity outcome occurring if an AGI passes the Turing test. Our question was broken into three parts, for each of three possible development scenarios: if the first AGI that can pass the Turing test is created by an open source project, the United States military, or a private company focused on commercial profit.

This set of questions marked another instance in which the experts wildly lacked consensus. Four experts estimated a less than 20% chance of a negative outcome, regardless of the development scenario. Four experts estimated a greater than 60% chance of a negative outcome, regardless of the development scenario. Only four experts gave the same estimate for all three development scenarios; the rest of the experts reported different estimates of which development scenarios were more likely to bring a negative outcome. Several experts were more concerned about the risk from AGI itself, whereas others were more concerned that AGI could be misused by humans who controlled it.

Some interesting insights can be found in the experts' orderings of the riskiness of the development scenarios. Of the 11 experts who gave different dangerousness estimates for each of the three scenarios, 10 gave the private company scenario the middle value. Of these 10, 6 gave the US military scenario the highest value and 4 gave it the lowest value. Thus the open source scenario and the US military scenario tend to be perceived opposites in terms of danger – but experts are divided on which is the safe one of the pair! Experts who estimated that the US military scenario is relatively safe noted that the US military faces strong moral constraints, has experience handling security issues, and is very reluctant to develop technologies that may backfire (such as biological weapons), whereas open source development lacks these features and cannot easily prevent the "deliberate insertion of malicious code". In contrast, experts who estimated that the open source scenario is relatively safe praised the transparency of open source development and its capacity to bring more minds to the appropriate problems, and felt the military has a tendency to be destructive.

Several experts noted potential impacts of AGI other than the catastrophic. One predicted that "in thirty years, it is likely that virtually all the intellectual work that is done by trained human beings such as doctors, lawyers, scientists, or programmers, can be done by computers for pennies an hour. It is also likely that with AGI the cost of capable robots will drop, drastically decreasing the value of physical labor. Thus, AGI is likely to eliminate almost all of today's decently paying jobs." This would be disruptive, but not necessarily bad. Another expert thought that, "societies could accept and promote the idea that AGI is mankind's greatest invention, providing great wealth, great health, and early access to a long and pleasant retirement for everyone." Indeed, the experts' comments suggested that the potential for this sort of positive outcome is a core motivator for much AGI research.

Conclusions of the Survey

In the broadest of terms, our results concur with those of the two previous studies mentioned above. All three studies suggest that significant numbers of interested, informed individuals believe it is likely that AGI at the human level or beyond will occur around the middle of this century, and plausibly even sooner. Due to the greater depth of the questions, our survey also revealed some interesting additional information, such as the disagreement among experts over the likely order of AGI milestones and the relative safety of different AGI development scenarios. The experts' suggestions regarding funding (with several advocating modest funding increases distributed among a broad range of research groups) are also potentially valuable.

6

What Is General Intelligence?

AGI researchers Shane Legg and Marcus Hutter wrote a paper in 2007 collecting 70+ different definitions of intelligence from the scientific literature The ten definitions that most reflect a similar perspective to Legg and Hutter's perspective, and ours, are as follows:

1. "It seems to us that in intelligence there is a fundamental faculty, the alteration or the lack of which, is of the utmost importance for practical life. This faculty is judgment, otherwise called good sense, practical sense, initiative, the faculty of adapting oneself to circumstances." A. Binet
2. "The capacity to learn or to profit by experience."...
3. "Ability to adapt oneself adequately to relatively new situations in life." R. Pinter
4. "A person possesses intelligence insofar as he has learned, or can learn, to adjust himself to his environment." S. S. Colvin
5. "We shall use the term 'intelligence' to mean the ability of an organism to solve new problems" W. V. Bingham
6. "A global concept that involves an individual's ability to act purposefully, think rationally, and deal effectively with the environment." D. Wechsler
7. "Individuals differ from one another in their ability to understand complex ideas, to adapt effectively to the environment, to learn from experience, to engage in various forms of reasoning, to overcome obstacles by taking thought." American Psychological Association
8. "I prefer to refer to it as 'successful intelligence.' And the reason is that the emphasis is on the use of your intelligence to achieve success in your life. So I define it

as your skill in achieving whatever it is you want to attain in your life within your sociocultural context — meaning that people have different goals for themselves, and for some it's to get very good grades in school and to do well on tests, and for others it might be to become a very good basketball player or actress or musician." R. J. Sternberg

9. "Intelligence is part of the internal environment that shows through at the interface between person and external environment as a function of cognitive task demands." R. E. Snow
10. "Certain set of cognitive capacities that enable an individual to adapt and thrive in any given environment they find themselves in, and those cognitive capacities include things like memory and retrieval, and problem solving and so forth. There's a cluster of cognitive abilities that lead to successful adaptation to a wide range of environments." D. K. Simonton

All these variants roughly say the same sort of thing, but different researchers have different slants. The most common feature of these definitions is an emphasis on learning, adaptability, and interaction with some external environment. Intelligence does not exist separately from the ability to interact with an environment, but rather is developed through it. In the view represented here, there is no magical higher order process that establishes intelligence independent of experience. A truly intelligent system is able to observe the environment, identify the problem, formulate goals, and then devise, implement, and test a solution. Only being able to reason about a previously defined domain of situations, problems, and solutions – such as a chess computer does – is not such a powerful kind of intelligence, in this sense.

As Legg and Hutter put it: "intelligence is not the ability to deal with a fully known environment, but rather the ability to deal with some range of possibilities which cannot be wholly anticipated." In other words, the definition we formulated at the outset: The ability to autonomously achieve complex goals in complex, changing environments.

Note that nothing in these definitions makes any statements about emotions. Intelligence does not require the ability to love, anger, or be bored. Also, the ability to observe a situation is intentionally stated vaguely. No specific sense – vision, audition, etc. – is specifically required in order to be generally intelligent. A human-like body is not required, nor is any kind of robot body nor virtual body, etc. This kind of sensorimotor apparatus may be very valuable for developing and maintaining some particular sorts of intelligence, including some of value and relevant to humans, but it's not critical to intelligence generally conceived. Potentially one could have an intelligence whose only "world" was what it saw through a Web browser, for example. This would not be a very human-like intelligence, but it might be human-level and perhaps even superhuman in some ways.

In general, the connection between the abstract nature of intelligence and the particularities of human intelligence is something that's understood only broadly and conceptually, at present. A full understanding of this connection probably awaits much fuller development of both AGI and cognitive neuroscience – though I've done some thinking in this direction, and will share some of my thoughts below.

The lack of a broadly agreed-upon definition for the "I" in "AI" has long been the source of sardonic humor among AI researchers – and criticism from those outside the field But I have always thought this criticism largely misplaced. After all, biology lacks a crisp definition of "life", and physics lacks crisp definitions of "space", "time" or "cause." So what? Sometimes foundational concepts like these elude precise definition, but serve to spawn other more crisply definable and measurable concepts.

One humorous definition goes: "Intelligence is whatever humans can do that computers can't do yet." There is some truth to this quip: for instance, most people would say that playing chess requires intelligence. Yet, now that we see Deep Blue playing grandmaster-beating chess via some simple mathematical algorithms (rather than through any human-like intuition or strategy), there is a temptation to reclassify chess-playing as something on the borderline of intelligence, rather than something definitively involving intelligence.

A less sarcastic slant on the same basic observation would be to say that AI research has helped us to clarify our understanding of what intelligence really is. The fact that various activities (like chess, navigation, medical diagnosis, or Web search), apparently requiring considerable general intelligence in humans, can be carried out by simple specialized algorithms – this fact was not at all **obvious** from the get-go, and constitutes a significant discovery on the part of the AI field. And the fact that similar simple algorithms **cannot** be used to integrate a broad understanding of real-world context into the solution of particular problems, also was not initially obvious, and has been discovered by the AI field during the last 5 decades. So, as well as creating a lot of very useful specialized AI programs, the corpus of AI work done so far has taught us a great deal about the nature of intelligence and its relationship with computer technology. Among many other things, we have learned that a key focus area for AGI research should be integration of contextual knowledge into problem-solving, and that focusing on solving narrowly-defined problems is not the right way to approach general-purpose AI.

A few prescient researchers limned some of these lessons, at least in sketchy form, even in the early days of the field. In Alan Turing's classic paper from the 1950s, the computing pioneer proposed what is now called the Turing test – a way of gauging whether one has created an AI with human-level intelligence. The Turing test basically says: "Write a computer program that can simulate a human in a text-based conversational interchange—and any human should be willing to consider it intelligent." The Turing test certainly has its weaknesses – but it does place the focus on something broad-based and requiring conceptual understanding, rather than a specific task or problem. Turing understood that the crux of human intelligence lies, not in the specialized tasks that certain clever humans do so much better than others, but rather in the commonsense processes of reasoning, learning, communicating and doing that we all carry out every day.

Indeed, if an AI can chat with educated humans for a couple hours, and fool them into think it's human, then I'm comfortable saying that AI really does have human-level intelligence. That is: I do buy the Turing Test as a "sufficient condition" for general intelligence. However, I don't buy it as a "necessary condition." It seems quite possible that a radically non-human intelligent computer system could develop, as unable to imitate humans as we are to imitate dogs or anteaters or sharks, but still just as intelligent as humans, or even more so.

Also, partial success at tricking humans into believing an AI is human, seems a particularly poor measure of partial progress toward human level AI – a point that I'll enlarge on later on, in reviewing some conversations with "chat bots" that try to imitate human conversation without any real understanding. So this kind of trickery does not seem a good way to make a graded AI-Q score, with current AIs near the bottom and human-level general intelligence further up. But even outside the scope of Turing Test style conversation, this kind of graded AI-Q test seems extremely difficult to come by, given our present level of understanding of AGI.

Overall, I think I have a fairly good *theoretical* idea of what intelligence is all about – beginning from the basic idea of "achieving complex goals in complex environments", mentioned above – and I'll say more about that in a moment. But neither I nor anyone else has a really good IQ test for would-be AGIs. The creation of such a test might be an interesting task, but I suspect it can't be effectively approached until we have a population of fairly similar, fairly advanced AGI systems – which we can then study, to obtain data to guide the creation of AGI intelligence metrics. Human IQ tests will probably guide the development of such tests only loosely. After all, they work moderately well within a single culture, but much worse across the spectrum of human cultures — so how much worse will they work across species, or across different types of computer programs, which may well be as different as different species of animals?

Defining General Intelligence in Theory and Practice

The same pair of researchers who collected the 70+ definitions of intelligence mentioned above – Shane Legg and Marcus Hutter – also wrote a paper presenting a formal mathematical theory of general intelligence. The equations are moderately hairy but the basic idea is pretty simple. They model an environment as something that interacts with an intelligent system, giving it percepts and rewards and accepting actions from it. And they say that a system's intelligence is its ability at getting rewarded, averaged over all possible environments. The trick is that the average is "weighted" by complexity – so that getting rewarded in a complex environment is weighted more than getting rewarded in a simple one. The trickier trick is that there's no objective measure of complexity – so that their definition is relative to which complexity measure you use.[4]

These are fairly obvious concepts, and a lot of researchers said similar things previously (for instance, I did in my 1993 book *The Structure of Intelligence*, and Ray Solomonoff did back in the 1960s). But Legg and Hutter crossed the t's and dotted the i's – and, leveraging Hutter's earlier work, they proved the existence

---

[4] Formally speaking relative to an assumed Universal Turing Machine. Also, for the math geeks in the audience: their definition assumes an exponential decaying complexity measure over the space of environments, which causes the most complex environments to be weighted far more than the vast bulk of other ones, a property that may not be desirable.

of a computer program displaying very high intelligence according to their definition. The only catch is that this program, called AIXI, requires an infinite amount of computing power to run. There are also some close approximations to AIXI that can use a finite amount of computing power – but still are totally impractical to build in reality. Marcus Hutter and a number of other researchers are now working on more practical approximations to AIXI – an effort that so far has led to some interesting algorithms and theories, but no pragmatic breakthroughs in terms of building advanced AGI systems.

I wrote a paper in 2010 that generalizes their definition a bit – extending the kind of complexity weighting involved, optionally incorporating the amount of computational effort required, and looking at systems that try to achieve goals represented in their minds rather than trying to get reward from the environment. I think these extensions have some technical value, but they don't radically change the picture. They share the main weakness of Legg and Hutter's approach – which is being very mathematical, and somewhat disconnected from the practicalities of real-world intelligent systems.

Human beings and other animals are simply not very intelligent according to this kind of abstract definition. If you put us in an arbitrary, highly complex environment – chosen from the scope of all mathematically possible environments – then we probably won't do anything very smart, and won't lead a very rewarding existence. In principle, perhaps, we could cope with any sort of complex environment you stick us in – if we have enough time and resources. For example, we could potentially build an AGI to figure out how to cope with the environment and help us out! But in reality, given finite amounts of time and space, we would screw up royally in any environment not fairly similar to the ones that we've evolved to deal with.

In other words, we're not so intelligent in terms of average behavior over all mathematically possible environments – rather, we're fairly intelligent in the context of environments of the general type that we've evolved for. We're land animals and we're pretty good at managing situations involving piecing solid objects together – but put us underwater where it's all about subtle interactions between fluid flows, and we become a lot stupider. 30% of our brains are devoted to vision processing, so if you put us in a dark place where sonar is the best way to get around, we become a lot dumber. We're so terrible at mathematical calculations, that we focus on areas of math where we can use analogies to things like vision (e.g. geometry) and language (e.g. logic, algebra) that we are better evolved to think about. We evolved to handle social situations involving the <300 people in a Stone Age tribe, so we get easily bollixed by the larger, more diffuse social networks that modern society presents us with. And so forth.

All this may make you wonder just how "general" is our intelligence after all!! But still, there's something to the fact that, in principle, given enough time and resources, we have the flexibility of mind to solve just about any problem. And there's no doubt that our intelligence is dramatically more general than that of Deep Blue, or a mouse.

Harking back to the formal theory of general intelligence, one can quantify the "generality" of an intelligent system by the entropy (the breadth or spread) of the class of environments in which it's good at achieving goals. A system that's really good at achieving goals in a narrow class of environments, maybe smart, but doesn't have so much generality of intelligence. A system that's moderately good at achieving goals in a broad variety of environment, may be said to have more generality of intelligence. Humans have both more intelligence, and more generality of intelligence, than Deep Blue or a mouse! A mouse has way more generality of intelligence than Deep Blue, even though mice are lousy chess players.

You might also question all this emphasis on rewards and goals. People do all sorts of things, some of which seem to have little to do with achieving goals, and some of which seem to have little to do with achieving near or long term rewards. We're complex self-organizing systems displaying all manner of different behaviors. Goals are things we sometimes try to follow, or ascribe to each other descriptively, rather than things that rigorously govern all our behaviors. And while sometimes we seek short or long term rewards, we also engage in a lot of activities that seem to have little reward at all, for ourselves or anybody. But, if you want to measure or define a system's intelligence, it still seems the best way to do this is relative to some collection of goals. Even if not everything a system does is intelligent, you can still measure its intelligence via the complexity of the goals it can achieve in complex environments. Otherwise, it's hard to see how to distinguish intelligence from more general "self-organizing complexity, coupled with the environment" – which is also interesting and important, but a broader sort of concept.

So, when we talk about building an artificial general intelligence in practice, what we're mostly talking about is **building a system that in principle could achieve nearly any goal in any environment given sufficient time and resources; and that in practice is good at achieving complex goals similar to those goals needed for the survival and flourishing of a human-like organism in a human society, in environments like the ones humans lived in historically or live in presently**. That's not very compact or elegant, but it's what AGI in reality is mainly about. Eventually the AGI field may branch beyond this, and get concerned with building broader kinds of AGI systems displaying varieties of intelligence further detached from human history and perhaps even incomprehensible to the human mind.

Viewed in this way, the practical AGI task has a lot to do with the specifics of human bodies, societies and environments – which is somewhat messy and not very mathematical, but seems to be the reality of the matter. The ideal theory of AGI, in my view, would be something like a recipe for producing intelligent systems from environments. You would feed the theory a description of the environment and goals that an intelligent system needs to deal with – and the theory would spit out a description of what kind of system could be intelligent with respect to that set of environment and goals, given limited computing resources. Unfortunately we don't have this kind of theory right now – I've spent some time trying to formulate it, but I don't feel particularly close to success at the moment. So we need to proceed in this direction via a more intuitive approach: given an intuitive, cross-disciplinary understanding of the environment and goals pertinent to a human-level, roughly human-like AGI, integrate all the knowledge we have from various sources to figure out how to craft an AGI capable of dealing with these environments and goals.

As a single, important example, consider some of the main types of memory that cognitive psychologists have identified as being critical to human intelligence, and the operation of the human brain:

- **Declarative** memory, for facts and beliefs
- **Procedural** memory, for practical, routinized knowledge of how to do stuff (the way we can know how to serve a tennis ball, or how to seduce a girl, or how to prove a theorem, even though we can't accurately describe in language exactly what we do to accomplish these things)
- **Sensorimotor** memory, recalling what we perceive and enact
- **Episodic** memory, our life-history memory of the stories we've been through

Why are these types of memory important for human intelligence and human-like AGI? Because of the kinds of goals and environments humans have. Our bodies imply an obvious need for sensorimotor memory. Our social lives imply an obvious need for episodic memory. Our facility for and habit of linguistic communication implies a need for declarative memory. Our ability to learn via practice and communicate via example implies a need for procedural memory. The particularities of our embodiment and environment leads fairly clearly to the need for the various types of memory that the human brain possesses. And then each of these types of memory must be served by appropriate learning mechanisms – a train of thought that has significant implications for AGI design. But the main point for now is simply that the specifics of an intelligent system's cognitive architecture, connect fairly closely to the specifics of the environment and goals that the system is intended to primarily deal with. This is how the abstract notion of intelligence as "achieving complex goals in complex environments" connects with the reality of real-world intelligent systems doing real things.

7

Why an Intelligence Explosion is Probable

The path from here to vastly superhuman AGI, I've suggested, can meaningfully be broken down into two phases: Phase 1, the creation of human-level AGI; Phase 2, the path from human-level AGI into the unknown realm beyond. The second of these phases is what I.J. Good referred to as the "intelligence explosion," in his famous 1965 quote:

> *Let an ultra-intelligent machine be defined as a machine that can far surpass all the intellectual activities of any man however clever. Since the design of machines is one of these intellectual activities, an ultra-intelligent machine could design even better machines; there would then unquestionably be an 'intelligence explosion,' and the intelligence of man would be left far behind. Thus the first ultra-intelligent machine is the last invention that man need ever make.*

Some would argue against the wisdom of isolating this kind of "intelligence explosion" as an event unto itself, preferring to emphasize the continuity of such a potential explosion, not only with the creation of superhuman AGI, but also with the larger, longer "explosion" of intelligence on Earth, beginning with the origins of life (or maybe even before). But while I think this broad-scope view is valuable, I also think it's important to recognize the singular nature of the intelligence explosion a la I.J. Good. Engineering is different from evolution, even though it was invented by evolved beings in the indirect pursuit of their evolutionary goals; and the recursive self-improvement of engineered minds, if it does indeed occur, will have different characteristics than the natural selection driven evolution of intelligence that preceded it.

My own research work is aimed at carrying off Phase 1, "the creation of human-level AGI", in a way that will not only lead to beneficial results straight away, but also bias the odds toward a positive version of Phase 2, Good's intelligence explosion. To me, it's always seemed fairly intuitively obvious that, once human-level AGI is achieved, a Good-style intelligence explosion will follow. Whether this "explosion" will take months (seems unlikely), years (seems probable), decades (maybe), or centuries (IMO, almost surely not) is less obvious. But the potential of a digitally-implemented, engineered human-level AGI to analyze, refine and improve itself, in the a manner leading to exponential improvement in its general intelligence, seems plain. The human brain is hard to study and hard to modify, because it wasn't created with analysis and improvement in mind. Engineered systems can be instrumented differently, leading to much more straightforward routes to improvement. No doubt the path from human-level AGI to dramatically superhuman AGI will involve many challenges, but, these will surely be qualitatively different ones from those involved with understanding and improving human brains, or with achieving human-level AGI in the first place.

But as obvious as Good's view seems to me and some others, unsurprisingly, not all futurist thinkers agree. Skeptics often cite limiting factors that could stop an intelligence explosion from happening, even after the achievement of human-level AGI; and in a 2010 post on the Extropy email discussion list, the futurist Anders Sandberg articulated some of those possible limiting factors, in a particularly clear way:

> *One of the things that struck me during our Winter Intelligence workshop on intelligence explosions was how confident some people were about the speed of recursive self-improvement of AIs, brain emulation collectives or economies. Some thought it was going to be fast in comparison to societal adaptation and development timescales (creating a winner takes all situation), some thought it would be slow enough for multiple superintelligent*

agents to emerge. This issue is at the root of many key questions about the singularity (One superintelligence or many? How much does friendliness matter?).

It would be interesting to hear this list's take on it: what do you think is the key limiting factor for how fast intelligence can amplify itself?

1. *Economic growth rate*
2. *Investment availability*
3. *Gathering of empirical information (experimentation, interacting with an environment)*
4. *Software complexity*
5. *Hardware demands vs. available hardware*
6. *Bandwidth*
7. *Lightspeed lags*

Clearly many more can be suggested. But which bottlenecks are the most limiting, and how can this be ascertained?"

Following Anders' post presenting these issues, the AGI researcher Richard Loosemore posted a detailed reply explaining why he thought none of these were such serious objections. I liked Richard's reply so I asked him if I could clean it up a bit and turn it into an article for H+ Magazine – he graciously agreed, and then it turned into more of a collaborative writing project, resulting in a jointly authored article with the title "Why An Intelligence Explosion is Probable," elaborating considerably on Richard's original email.

And so, here is the article that Richard and I wrote…

Why an Intelligence Explosion is Probable

by Richard Loosemore and Ben Goertzel

[This chapter originally appeared in the form of an H+ Magazine article, which began with a list of Anders' Sandberg's objections to the intelligence explosion idea, as cited just above.]

We are grateful to Sandberg for presenting this list of questions because it makes it especially straightforward for us to provide a clear refutation, in this article, of the case against the viability of an intelligence explosion. We intend to explain why these bottlenecks are unlikely to be significant issues, and thus why, as I.J. Good predicted, an intelligence explosion is indeed a very likely outcome.

The One Clear Prerequisite for an Intelligence Explosion

To begin, we need to delimit the scope and background assumptions of our argument. In particular, it is important to specify what kind of intelligent system would be capable of generating an intelligence explosion.

According to our interpretation, there is one absolute prerequisite for an explosion to occur, and that is that an artificial general intelligence (AGI) must become smart enough to understand its own design. In fact, by choosing to label it an "artificial general intelligence" we have already said, implicitly, that it will be capable of self-understanding, since the definition of an AGI is that it has a broad set of intellectual capabilities that include all the forms of intelligence that we humans possess—and at least some humans, at that point, would be able to understand AGI design.

But even among humans there are variations in skill level and knowledge, so the AGI that triggers the explosion must have a sufficiently advanced intelligence that it can think analytically and imaginatively about how to manipulate and improve the design of intelligent systems. It is possible that not all humans are able to do this, so an AGI that met the bare minimum requirements for AGI-hood—say, a system smart enough to be a general household factotum—would not necessarily have the ability to work in an AGI research laboratory. Without an advanced AGI of the latter sort, there would be no explosion, just growth as usual, because the rate-limiting step would still be the depth and speed at which humans can think.

The sort of fully-capable AGI we're referring to might be called a "seed AGI", but we prefer to use the less dramatic phrase "self-understanding, human-level AGI." This term, though accurate, is still rather cumbersome, so we will sometimes use the phrase "the first real AGI" or just "the first AGI" to denote the same idea. In effect, we are taking the position that for something to be a proper artificial general intelligence it has to be capable of competing with the best that the human intellect can achieve, rather than being limited to a bare minimum. So the "first AGI" would be capable of initiating an intelligence explosion.

Distinguishing the Explosion from the Build-Up

Given that the essential prerequisite for an explosion to begin would be the availability of the first self-understanding, human-level AGI, does it make sense to talk about the period leading up to that arrival—the period during which that first real AGI was being developed and trained—as part of the intelligence explosion proper? We would argue that this is not appropriate, and that the true start of the explosion period should be considered to be the moment when a sufficiently well qualified AGI turns up for work at an AGI research laboratory. This may be different from the way some others use the term, but it seems consistent with I.J. Good's original usage. So our concern here is to argue for the high probability of an intelligence explosion, given the assumption that a self-understanding, human-level AGI has been created.

By enforcing this distinction, we are trying to avoid possible confusion with the parallel (and extensive!) debate about whether a self-understanding, human-level AGI can be built at all. Questions about whether an AGI with "seed level capability" can plausibly be constructed, or how long it might take to arrive, are of course quite different. A spectrum of opinions on this issue, from a survey of AGI researchers at a 2009 AGI conference, were presented in a 2010 H+ magazine article[5]. In that survey, of an admittedly biased sample, a majority felt that an AGI with this capability could be achieved by the middle of this century, though a substantial plurality felt it was likely to happen much further out. While we have no shortage of our own thoughts and arguments on this matter, we will leave them aside for the purpose of the present paper.

What Constitutes an "Explosion"?

How big and how long and how fast would the explosion have to be to count as an "explosion"?

Good's original notion had more to do with the explosion's beginning than its end, or its extent, or the speed of its middle or later phases. His point was that in a short space of time a human-level AGI would probably explode into a significantly transhuman AGI, but he did not try to argue that subsequent improvements would continue without limit. We, like Good, are primarily interested in the explosion from human-level AGI to an AGI with, very loosely speaking, a level of general intelligence 2-3 orders of magnitude greater than the human level (say, 100H or 1,000H, using 1H to denote human-level general intelligence). This is not because we are necessarily skeptical of the explosion

[5] http://hplusmagazine.com/2010/02/05/how-long-till-human-level-ai/

continuing beyond such a point, but rather because pursuing the notion beyond that seems a stretch of humanity's current intellectual framework.

Our reasoning, here, is that if an AGI were to increase its capacity to carry out scientific and technological research, to such a degree that it was discovering new knowledge and inventions at a rate 100 or 1,000 times the rate at which humans now do those things, we would find that kind of world unimaginably more intense than any future in which humans were doing the inventing. In a 1,000H world, AGI scientists could go from high-school knowledge of physics to the invention of relativity in a single day (assuming, for the moment, that the factor of 1,000 was all in the speed of thought—an assumption we will examine in more detail later). That kind of scenario is dramatically different from a world of purely human inventiveness — no matter how far humans might improve themselves in the future, without AGI, its seems unlikely there will ever be a time when a future Einstein would wake up one morning with a child's knowledge of science and then go on to conceive the theory of relativity by the following day—so it seems safe to call that an "intelligence explosion."

This still leaves the question of how *fast* it has to arrive, to be considered explosive. Would it be enough for the first AGI to go from 1H to 1,000H in the course of a century, or does it have to happen much quicker, to qualify?

Perhaps there is no need to rush to judgment on this point. Even a century-long climb up to the 1,000H level would mean that the world would be very different for the rest of history. The simplest position to take, we suggest, is that if the human species can get to the point where it is creating new types of intelligence that are themselves creating intelligences of greater power, then this is something new in the world (because at the moment all we can do is create human babies of power 1H), so even if this process happened rather slowly, it would still be an explosion of sorts. It might not be a Big Bang, but it would at least be a period of Inflation, and both could eventually lead to a 1,000H world.

Defining Intelligence (Or Not)

To talk about an intelligence explosion, one has to know what one means by "intelligence" as well as by "explosion". So it's worth reflecting that there are currently no measures of general intelligence that are precise, objectively defined and broadly extensible beyond the human scope.

However, since "intelligence explosion" is a qualitative concept, we believe the commonsense qualitative understanding of intelligence suffices. We can address Sandberg's potential bottlenecks in some detail without needing a precise measure, and we believe that little is lost by avoiding the issue. We will say that an intelligence explosion is something with the potential to create AGI systems as far beyond humans as humans are beyond mice or cockroaches, but we will not try to pin down exactly how far away the mice and cockroaches really are.

Key Properties of the Intelligence Explosion

Before we get into a detailed analysis of the specific factors on Sandberg's list, some general comments are in order.

Inherent Uncertainty. Although we can try our best to understand how an intelligence explosion might happen, the truth is that there are too many interactions between the factors for any kind of reliable conclusion to be reached. This is a complex-system interaction in which even the tiniest, least-anticipated factor may turn out to be either the rate-limiting step or the spark that starts the fire. So there is an irreducible uncertainty involved here, and we should be wary of promoting conclusions that seem too firm.

General versus Special Arguments. There are two ways to address the question of whether or not an intelligence explosion is likely to occur. One is based on quite general considerations. The other involves looking at specific pathways to AGI. An AGI researcher (such as either of the authors) might believe they understand a great deal of the technical work that needs to be done to create an intelligence explosion, so they may be confident of the plausibility of the idea for that reason alone. We will restrict ourselves here to the first kind of argument, which is easier to make in a relatively non-controversial way, and leave aside any factors that might arise from our own understanding about how to build an AGI.

The "Bruce Wayne" Scenario. When the first self-understanding, human-level AGI system is built, it is unlikely to be the creation of a lone inventor working in a shed at the bottom of the garden, who manages to produce the finished product without telling anyone. Very few of the "lone inventor" (or "Bruce Wayne") scenarios seem plausible. As communication technology advances and causes cultural shifts, technological progress is increasingly tied to rapid communication of information between various parties. It is unlikely that a single inventor would be able to dramatically outpace multi-person teams working on similar projects; and also unlikely that a multi-person team would successfully keep such a difficult and time-consuming project secret, given the nature of modern technology culture.

Unrecognized Invention. It also seems quite implausible that the invention of a human-level, self-understanding AGI would be followed by a period in which the invention just sits on a shelf with nobody bothering to pick it up. The AGI situation would probably not resemble the early reception of inventions like the telephone or phonograph, where the full potential of the invention was largely unrecognized. We live in an era in which practically-demonstrated technological advances are broadly and enthusiastically communicated, and receive ample investment of dollars and expertise. AGI receives relatively little funding now, for a combination of reasons, but it is implausible to expect this situation to continue in the scenario where highly technically capable human-level AGI systems exist. This pertains directly to the economic objections on Sandberg's list, as we will elaborate below.

Hardware Requirements. When the first human-level AGI is developed, it will either require a supercomputer-level of hardware resources, or it will be achievable with much less. This is an important dichotomy to consider, because world-class supercomputer hardware is not something that can quickly be duplicated on a large scale. We could make perhaps hundreds of such machines, with a massive effort, but probably not a million of them in a couple of years.

Smarter versus Faster. There are two possible types of intelligence speedup: one due to faster operation of an intelligent system (clock speed increase) and one due to an improvement in the type of mechanisms that implement the thought processes ("depth of thought" increase). Obviously both could occur at once (and there may be significant synergies), but the latter is ostensibly more difficult to achieve, and may be subject to fundamental limits that we do not understand. Speeding up the hardware, on the other hand, is something that has been going on for a long time and is more mundane and reliable. Notice that both routes lead to greater "intelligence," because even a human level of thinking and creativity would be more effective if it were happening a thousand times faster than it does now.

It seems quite possible that the general class of AGI systems can be architected to take better advantage of improved hardware than would be the case with intelligent systems very narrowly imitative of the human brain. But even if this is not the case, brute hardware speedup can still yield dramatic intelligent improvement.

Public Perception. The way an intelligence explosion presents itself to human society will depend strongly on the rate of the explosion in the period shortly after the development of the first self-understanding human-level AGI. For instance, if the first such AGI takes five years to "double" its intelligence, this is a very different matter than if it takes two months. A five-year time frame could easily arise, for example, if the first AGI required an extremely expensive supercomputer based on unusual hardware, and the owners of this hardware were to move slowly. On the other hand, a two-month time frame could more easily arise if the initial AGI were created using open source software and commodity hardware, so that a doubling of intelligence only required addition of more hardware and a modest number of software changes. In the former case, there would be more time for governments, corporations and individuals to adapt to the reality of the intelligence explosion before it reached dramatically transhuman levels of intelligence. In the latter case, the intelligence explosion would strike the human race more suddenly. But this potentially large difference in human perception of the events would correspond to a fairly minor difference in terms of the underlying processes driving the intelligence explosion.

So – now, finally, with all the preliminaries behind us, we will move on to deal with the specific factors on Sandberg's list, one by one, explaining in simple terms why each is not actually likely to be a significant bottleneck. There is much more that could be said about each of these, but our aim here is to lay out the main points in a compact way.

Objection 1: Economic Growth Rate and Investment Availability

The arrival, or imminent arrival, of human-level, self-understanding AGI systems would clearly have dramatic implications for the world economy. It seems inevitable that these dramatic implications would be sufficient to offset any factors related to the economic growth rate at the time that AGI began to appear. Assuming the continued existence of technologically advanced nations with operational technology R&D sectors, if self-understanding human-level AGI is created, then it will almost surely receive significant investment. Japan's economic growth rate, for example, is at the present time somewhat stagnant, but there can be no doubt that if any kind of powerful AGI were demonstrated, significant Japanese government and corporate funding would be put into its further development.

And even if it were not for the normal economic pressure to exploit the technology, international competitiveness would undoubtedly play a strong role. If a working AGI prototype were to approach the level at which an explosion seemed possible, governments around the world would recognize that this was a critically important technology, and no effort would be spared to produce the first fully-functional AGI "before the other side does." Entire national economies might well be sublimated to the goal of developing the first superintelligent machine, in the manner of Project Apollo in the 1960s. Far from influencing the intelligence explosion, economic growth rate would be *defined* by the various AGI projects taking place around the world.

Furthermore, it seems likely that once a human-level AGI has been achieved, it will have a substantial—and immediate—practical impact on multiple industries. If an AGI could understand its own design, it could also understand and improve other computer software, and so have a revolutionary impact on the software industry. Since the majority of financial trading on the US markets is now driven by program trading systems, it is likely that such AGI technology would rapidly become indispensable to the finance industry (typically an early adopter of any software or AI innovations). Military and espionage establishments would very likely also find a host of practical applications for such technology. So, following the achievement of self-understanding, human-level AGI, and complementing the allocation of substantial research funding aimed at outpacing the competition in achieving ever-smarter AGI, there is a great likelihood of funding aimed at practical AGI applications, which would indirectly drive core AGI research along.

The details of how this development frenzy would play out are open to debate, but we can at least be sure that the economic growth rate and investment climate in the AGI development period would quickly become irrelevant.

However, there is one interesting question left open by these considerations. At the time of writing, AGI investment around the world is noticeably weak, compared with other classes of scientific and technological investment. Is it possible that this situation will continue indefinitely, causing so little progress to be made that no viable prototype systems are built, and no investors ever believe that a real AGI is feasible?

This is hard to gauge, but as AGI researchers ourselves, our (clearly biased) opinion is that a "permanent winter" scenario is too unstable to be believable. Because of premature claims made by AI researchers in the past, a barrier to investment clearly exists in the minds of today's investors and funding agencies, but the climate already seems to be changing. And even if this apparent thaw turns out to be illusory, we still find it hard to believe that there will not eventually be an AGI investment episode comparable to the one that kicked the internet into high gear in the late 1990s. Furthermore, due to technology advanced in allied fields (computer science, programming language, simulation environments, robotics, computer hardware, neuroscience, cognitive psychology, etc.), the amount of effort required to implement advanced AGI designs is steadily decreasing – so that as time goes on, the amount of investment required to get AGI to the explosion-enabling level will keep growing less and less.

Objection 2: Inherent Slowness of Experiments and Environmental Interaction

This possible limiting factor stems from the fact that any AGI capable of starting the intelligence explosion would need to do some experimentation and interaction with the environment in order to improve itself. For example, if it wanted to re-implement itself on faster hardware (most probably the quickest route to an intelligence increase) it would have to set up a hardware research laboratory and gather new scientific data by doing experiments, some of which might proceed slowly due to limitations of experimental technology.

The key question here is this: how much of the research can be sped up by throwing large amounts of intelligence at it? This is closely related to the problem of parallelizing a process (which is to say: You cannot make a baby nine times quicker by asking nine women to be pregnant for one month). Certain algorithmic problems are not easily solved more rapidly simply by adding more processing power, and in much the same way there might be certain crucial physical experiments that cannot be hastened by doing a parallel set of shorter experiments.

This is not a factor that we can understand fully ahead of time, because some experiments that look as though they require fundamentally slow physical processes—like waiting for a silicon crystal to grow, so we can study a chip fabrication mechanism—may actually be dependent on the intelligence of the experimenter, in ways that we cannot anticipate. It could be that instead of waiting for the chips to grow at their own speed, the AGI could do some clever micro-experiments that yield the same information faster.

The increasing amount of work being done on nanoscale engineering would seem to reinforce this point—many processes that are relatively slow today could be done radically faster using nanoscale solutions. And it is certainly feasible that advanced AGI could accelerate nanotechnology research, thus initiating a "virtuous cycle" where AGI and nanotech research respectively push each other forward (as foreseen by nanotech pioneer Josh Hall[6]). As current physics theory does not even rule out more

[6] http://hplusmagazine.com/2011/01/24/toward-intelligent-nano-factories-and-fogs/

outlandish possibilities like femtotechnology[7], it certainly does not suggest the existence of absolute physical limits on experimentation speed existing anywhere near the realm of contemporary science.

Clearly, there is significant uncertainty in regards to this aspect of future AGI development. One observation, however, seems to cut through much of the uncertainty. Of all the ingredients that determine how fast empirical scientific research can be carried out, we know that in today's world the intelligence and thinking speed of the scientists themselves must be one of the most important. Anyone involved with science and technology R&D would probably agree that in our present state of technological sophistication, advanced research projects are strongly limited by the availability and cost of intelligent and experienced scientists.

But if research labs around the world have stopped throwing more scientists at problems they want to solve, because the latter are unobtainable or too expensive, would it be likely that those research labs are also, quite independently, at the limit for the physical rate at which experiments can be carried out? It seems hard to believe that both of these limits would have been reached at the same time, because they do not seem to be independently optimizable. If the two factors of experiment speed and scientist availability could be independantly optimized, this would mean that even in a situation where there was a shortage of scientists, we could still be sure that we had discovered all of the fastest possible experimental techniques, with no room for inventing new, ingenious techniques that get over the physical-experiment-speed limits. In fact, however, we have every reason to believe that if we were to double the number of scientists on the planet at the moment, some of them would discover new ways to conduct experiments, exceeding some of the current

---

[7] http://hplusmagazine.com/2011/01/10/theres-plenty-more-room-bottom-beyond-nanotech-femtotech

speed limits. If that were not true, it would mean that we had quite coincidentally reached the limits of science talent and physical speed of data collecting at the same time—a coincidence that we do not find plausible.

This picture of the current situation seems consistent with anecdotal reports: Companies complain that research staff are expensive and in short supply; they do not complain that nature is just too slow. It seems generally accepted, in practice, that with the addition of more researchers to an area of inquiry, methods of speeding up and otherwise improving processes can be found.

So based on the actual practice of science and engineering today (as well as known physical theory), it seems most likely that any experiment – speed limits lie further up the road, out of sight. We have not reached them yet, and we lack any solid basis for speculation about exactly where they might be.

Overall, it seems we do not have concrete reasons to believe that this will be a fundamental limit that stops the intelligence explosion from taking an AGI from H (human-level general intelligence) to (say) 1,000 H. Increases in speed within that range (for computer hardware, for example) are already expected, even without large numbers of AGI systems helping out, so it would seem that physical limits, by themselves, would be very unlikely to stop an explosion from 1H to 1,000 H.

Objection 3: Software Complexity

This factor is about the complexity of the software that an AGI must develop in order to explode its intelligence. The premise behind this supposed bottleneck is that even an AGI with self-knowledge finds it hard to cope with the fabulous complexity of the problem of improving its own software.

This seems implausible as a limiting factor, because the AGI could always leave the software alone and develop faster hardware. So long as the AGI can find a substrate that gives it a thousand-fold increase in clock speed, we have the possibility for a significant intelligence explosion.

Arguing that software complexity will stop the *first* self-understanding, human-level AGI from being built is a different matter. It may stop an intelligence explosion from happening by stopping the precursor events, but we take that to be a different type of question. As we explained earlier, one premise of the present analysis is that an AGI can actually be built. It would take more space than is available here to properly address that question.

It furthermore seems likely that, if an AGI system is able to comprehend its own software as well as a human being can, it will be able to improve that software significantly beyond what humans have been able to do. This is because in many ways, digital computer infrastructure is more suitable to software development than the human brain's wetware. And AGI software may be able to interface directly with programming language interpreters, formal verification systems and other programming-related software, in ways that the human brain cannot. In that way the software complexity issues faced by human programmers would be significantly mitigated for human-level AGI systems. However, this is not a 100% critical point for our arguments, because even if software complexity remains a severe difficulty for a self-understanding, human-level AGI system, we can always fall back to arguments based on clock speed.

Objection 4: Hardware Requirements

We have already mentioned that much depends on whether the first AGI requires a large, world-class supercomputer, or whether it can be done on something much smaller.

This is something that could limit the initial speed of the explosion, because one of the critical factors would be the number of copies of the first AGI that can be created. Why would this be critical? Because the ability to *copy* the intelligence of a fully developed, experienced AGI is one of the most significant mechanisms at the core of an intelligence explosion. We cannot do this copying of adult, skilled humans, so human geniuses have to be rebuilt from scratch every generation. But if one AGI were to learn to be a world expert in some important field, it could be cloned any number of times to yield an instant community of collaborating experts.

However, if the first AGI had to be implemented on a supercomputer, that would make it hard to replicate the AGI on a huge scale, and the intelligence explosion would be slowed down because the replication rate would play a strong role in determining the intelligence-production rate.

However, as time went on, the rate of replication would grow, as hardware costs declined. This would mean that the rate of arrival of high-grade intelligence would increase in the years following the start of this process. That intelligence would then be used to improve the design of the AGIs (at the very least, increasing the rate of new-and-faster-hardware production), which would have a positive feedback effect on the intelligence production rate.

So if there was a supercomputer-hardware requirement for the first AGI, we would see this as something that would only dampen the initial stages of the explosion. Positive feedback after that would eventually lead to an explosion anyway.

If, on the other hand, the initial hardware requirements turn out to be modest (as they could very well be), the explosion would come out of the gate at full speed.

Objection 5: Bandwidth

In addition to the aforementioned cloning of adult AGIs, which would allow the multiplication of knowledge in ways not currently available in humans, there is also the fact that AGIs could communicate with one another using high-bandwidth channels. This is *inter-AGI bandwidth*, and it is one of the two types of bandwidth factors that could affect the intelligence explosion.

Quite apart from the communication speed between AGI systems, there might also be bandwidth limits inside a single AGI, which could make it difficult to augment the intelligence of a single system. This is *intra-AGI bandwidth*.

The first one—inter-AGI bandwidth—is unlikely to have a strong impact on an intelligence explosion because there are so many research issues that can be split into separately-addressable components. Bandwidth between the AGIs would only become apparent if we started to notice AGIs sitting around with no work to do on the intelligence amplification project, because they had reached an unavoidable stopping point and were waiting for other AGIs to get a free channel to talk to them. Given the number of different aspects of intelligence and computation that could be improved, this idea seems profoundly unlikely.

Intra-AGI bandwidth is another matter. One example of a situation in which internal bandwidth could be a limiting factor would be if the AGI's working memory capacity were dependent on the need for total connectivity—everything connected to everything else—in a critical component of the system. If this case, we might find that we could not boost working memory very much in an AGI because the bandwidth requirements would increase explosively. This kind of restriction on the design of working memory might have a significant effect on the system's depth of thought.

However, notice that such factors may not inhibit the initial phase of an explosion, because the clock speed, not the depth of thought, of the AGI may be improvable by several orders of magnitude before bandwidth limits kick in. The main element of the reasoning behind this is the observation that neural signal speed is so slow. If a brain-like AGI system (not necessarily a whole brain emulation, but just something that replicated the high-level functionality of the brain) could be built using components that kept the same type of processing demands, and the same signal speed as neurons, then we would be looking at a human-level AGI in which information packets were being exchanged once every millisecond. In that kind of system there would then be plenty of room to develop faster signal speeds and increase the intelligence of the system. The processing elements would also have to go faster, if they were not idling, but the point is that the bandwidth would not be the critical problem.

Objection 6: Lightspeed Lags

Here we need to consider the limits imposed by special relativity on the speed of information transmission in the physical universe. However, its implications in the context of AGI are not much different than those of bandwidth limits.

Lightspeed lags could be a significant problem if the components of the machine were physically so far apart that massive amounts of data (by assumption) were delivered with a significant delay. But they seem unlikely to be a problem in the initial few orders of magnitude of the explosion. Again, this argument derives from what we know about the brain. We know that the brain's hardware was chosen due to biochemical constraints. We are carbon-based, not silicon-and-copper-based, so there are no electronic chips in the head, only pipes filled with fluid and slow molecular gates in the walls of the pipes. But if nature was forced to use the pipes-and-ion-channels approach, that leaves us with plenty of scope for speeding things up using silicon and copper (and this is quite apart from all the other more exotic computing substrates that are now on the horizon). If we were simply to make a transition membrane depolarization waves to silicon and copper, and if this produced a 1,000x speedup (a conservative estimate, given the intrinsic difference between the two forms of signalling), this would be an explosion worthy of the name.

The main circumstance under which this reasoning would break down would be if, for some reason, the brain is limited on two fronts simultaneously: both by the carbon implementation and by the fact that other implementations of the same basic design are limited by disruptive light-speed delays. This would mean that all non-carbon-implementations of the brain take us up close to the lightspeed limit before we get much of a speedup over the brain. This would require a coincidence of limiting factors (two limiting factors just happening to kick in at exactly the same level), that we find quite implausible, because it would imply a rather bizarre situation in which evolution tried both the biological neuron design, and a silicon implementation of the same design, and after doing a side-by-side comparison of performance, chose the one that pushed the efficiency of all the information transmission mechanisms up to their end stops.

Objection 7: Human-Level Intelligence May Require Quantum (Or More Exotic) Computing

Finally we consider an objection not on Sandberg's list, but raised from time to time in the popular and even scientific literature. The working assumption of the vast majority of the contemporary AGI field is that human-level intelligence can eventually be implemented on digital computers, but the laws of physics as currently understood imply that, to simulate certain physical systems without dramatic slowdown, requires special physical systems called "quantum computers" rather than ordinary digital computers.

There is currently no evidence that the human brain is a system of this nature. Of course the brain has quantum mechanics at its underpinnings, but there is no evidence that it displays quantum coherence at the levels directly relevant to human intelligent behavior. In fact our current understanding of physics implies that this is unlikely, since quantum coherence has not yet been observed in any similarly large and "wet" system. Furthermore, even if the human brain were shown to rely to some extent on quantum computing, this wouldn't imply that quantum computing is necessary for human-level intelligence — there are often many different ways to solve the same algorithmic problem. And (the killer counterargument), even if quantum computing *were* necessary for human-level general intelligence, that would merely delay the intelligence explosion a little, while suitable quantum computing hardware was developed. Already the development of such hardware is the subject of intensive R&D.

Roger Penrose, Stuart Hameroff and a few others have argued that human intelligence may even rely on some form of "quantum gravity computing", going beyond what ordinary quantum computing is capable of. This is really a complete blue-sky speculation with no foundation in current science, so not worth discussing in detail; but broadly speaking, this claim may be treated according to the same arguments as we've presented above regarding quantum computing.

The Path from AGI to Intelligence Explosion Seems Clear

Summing up, then – the conclusion of our relatively detailed analysis of Sandberg's objections is that there is currently no good reason to believe that once a human-level AGI capable of understanding its own design is achieved, an intelligence explosion will fail to ensue.

The operative definition of "intelligence explosion" that we have assumed here involves an increase of the speed of thought (and perhaps also the "depth of thought") of about two or three orders of magnitude. If someone were to insist that a real intelligence explosion had to involve million-fold or trillion-fold increases in intelligence, we think that no amount of analysis, at this stage, could yield sensible conclusions. But since an AGI with intelligence = 1000 H might well cause the next thousand years of new science and technology to arrive in one year (assuming that the speed of physical experimentation did not become a significant factor within that range), it would be churlish, we think, not to call that an "explosion". An intelligence explosion of such magnitude would bring us into a domain that our current science, technology and conceptual framework are not equipped to deal with; so prediction beyond this stage is best done once the intelligence explosion has already progressed significantly.

Of course, even if the above analysis is correct, there is a great deal we do not understand about the intelligence explosion, and many of these particulars will remain opaque until we know precisely what sort of AGI system will launch the explosion. But our view is that the likelihood of transition from a self-understanding human-level AGI to an intelligence explosion should not presently be a subject of serious doubt.

8

Ten Common Objections to AGI – And Why They Don't Scare Me

(originally an H+ Magazine article)

I've heard a lot of supposed reasons why AGI can't be engineered, or why this won't happen soon. Some of them don't deserve to be dignified by any response. But some are reasonably sensible, even though I don't quite buy them.

This chapter gives a quick list of ten of the commoner objections I've heard, with each objection followed by my own -- highly opinionated and openly AGI-optimist – perspective. In the interest of concision and punchiness, I won't take many pains to explicitly justify my take on each objection here – the justifications are given elsewhere in the book!

1. Objection via Quantum Computing

The Claim: The brain is a quantum system, in the strong sense that its intelligent behaviors rely on macroscopic quantum coherence among its parts. If this is the case, emulating brain function on a classical digital computer would be extremely inefficient (though possible in principle).

My take: There's no evidence this is the case, and it would require some revolutionary changes in the science of macroscopic quantum coherence. However, this claim doesn't seem to violate known science in a totally egregious way.

But even if it's true, all this means is that we might need to engineer an AGI to run on quantum computers rather than digital computers... Which would delay the advent of AGI but not pose a fundamental obstacle.

Also note that, even if the brain uses quantum computing, that doesn't mean a human-level AGI also needs to. There are often many different ways to achieve the same engineering functions.

Note that the brain is well known to make use of various quantum phenomena in the lower levels of operation – like every other physical system. That's not the same as manifesting macroscopic quantum coherence in the way that would require use of a quantum computer for efficient brain emulation.

2. Objection via Hypercomputing

Even further out...

The Claim: The brain is a **hypercomputer**[8], which produces intelligent behaviors via computing functions that no conventional computer can compute! (Quantum computers can't do this, they can only compute certain functions **faster** than conventional computers.)

Roger Penrose and Stuart Hameroff [9] have famously hypothesized that the brain works using "quantum gravity computing" which makes use of hypercomputing. However, they haven't posited any specific quantum gravity theory that actually needs hypercomputing and explains any of the particulars of brain function.

My take: Not only is there no evidence for this objection, there are also conceptual problems regarding the very notion of evidence for hypercomputing.

Any finite set of finite-precision data (i.e. any scientific data, in the vein of current or prior understandings of science) can be explained by some computational model; furthermore, no specific hypercomputational model can ever be described in

---

[8] http://en.wikipedia.org/wiki/Hypercomputation
[9] http://www.quantumconsciousness.org/

detail in a finite list of words or mathematical symbols. Nevertheless, it's conceivable that some hypercomputational model could be used to qualitatively explain neural data, and that a community of people could collectively recognize it as the simplest explanation of that data (even though they couldn't communicate exactly why in finite series of linguistic or mathematical symbols). But this is pretty weird speculative territory, and there is not any evidence that anything like this could work. Right now the hypercomputable mind is just out-there speculation, independent of any specific, detailed analyses of specific neural or mental phenomena.

Clearly this sort of funky speculation can't be taken as a serious objection to engineering AGI!

3. Objection via the Specialness of Biology

The Claim: Even if AGI doesn't require quantum computing or hypercomputing, it requires a sufficiently ***special*** *computing infrastructure that it's not pragmatically feasible to achieve it using ordinary computing hardware (even if possible in principle according to computing theory). One really needs to use certain kinds of analogue or chemical computers to get human-level AGI with feasible efficiency.*

My take: It's possible. However, it's not the message you get from the bulk of the neuroscience literature today, and especially not from the computational neuroscience literature. Neuroscientists model neurons using differential equations models that run on ordinary computers; and other cells like glia also seem susceptible to such modeling. Henri Markram, Kwabena Boahen and many other computational neuroscientists are working on building bigger and bigger and better and better computational neuroscience models. Indeed, this whole branch of research might fail – science is never certain -- but at the moment it seems to be advancing quite successfully and rapidly (see a review of the literature that I co-authored[10]).

4. Objection via Complexity

The Claim: The human brain is a complex system, and each part is finely tuned to work with the other parts. There's no viable way to analytically determine how the behavior of the whole depends on the behaviors of the parts, because this was created via evolution-guided self-organization, which often tends to create messy complex systems that are hard to analyze precise. So a brain-like system is simply too complex (not just complicated, but complex in the sense of complex systems theory) to design and engineer using any of our commonly recognized engineering methodologies.

According to this view, detailed brain emulation could potentially work, but only if we had a good enough theory of brain and mind to know what level the emulation should focus on, so as to be able to intelligently tweak and test the system. Or, systems deviating further from brains could potentially lead to AGI, but only if we developed some radically new method of complex systems evolution/engineering, different from any known engineering methodology.

AGI researcher Richard Loosemore has championed this view on AGI email lists for some time now; see e.g. his Complex Cognitive Systems Manifesto[11].

My take: Yeah, the brain is complex, but not **that** complex. I think we will understand it analytically once we have better brain imaging data, enabled by better imaging tech. And, only some of that complexity is necessary for AGI, some of it is just the way

---

[10] http://www.informatik.uni-trier.de/~ley/db/journals/ijon/ijon74.html
[11] http://www.richardloosemore.com/papers

the brain happened to evolve. An AGI can be engineered to be more judicious in the complexity that it manifests.

However, the idea of new methods of creating complex systems, combining aspects of engineering and artificial evolution, seems interesting and may play a role in the creation of AGI somehow.

5. Objection via Difficulty and Lack of Resources

*The Claim: Even if none of the above arguments hold -- so that it's plausible to engineer an AGI via clever deployment of mathematics, science and engineering principles – AGI still probably won't happen just because it's **really hard** and society doesn't choose to allocate many resources toward this task.*

I think this is possible, but I'm working pretty hard to make it false. My thought is that once some AGI project demonstrates AGI with a certain threshold level of functionality, this will serve as a sort of "AGI Sputnik[12]" and get the world excited about devoting financial and attentional resources to AGI.

The question then becomes if it's too hard to get to that Sputnik level. I'm more optimistic than most AGI researchers, I admit: My own bet is that producing an AGI with the rough intelligence of a 3-4 year old human child is feasible within a 4-5 years of work by a team of 15 or so good AI programmers (see the OpenCog Roadmap[13]). The key of course is the right AGI architecture, and I believe the OpenCog system that I co-created is one viable option.

But setting aside my particular optimism about the potential of my own work, the broader question is whether **some** AGI-Sputnik type achievement is feasible given relatively limited resources – and note that the cost of computing hardware, and

---

[12] http://hplusmagazine.com/2011/03/30/seeking-the-sputnik-of-agi/

[13] http://opencog.org/roadmap/

the ease of producing complex software, are improving rapidly year on year. If so, then this objection fails.

6.Objection via Vagueness of the Goal

The Claim: The scientific community has no clear, commonly accepted definition of what "Human-level AGI" or "Transhuman AGI" really means. How can people build something if they don't know what they're building?

My take: "Intelligence" is a fuzzy, natural language term, without a single clear meaning. There is a mathematical definition of general intelligence[14], posited by Shane Legg and Marcus Hutter, which I've extended into a more practical formal theory of general intelligence[15]. Not everyone in the AGI community accepts this definition as the best one, but so what? The point is that the AGI field does have concrete formalizations of its goals.

And furthermore, pragmatic goals may be more useful than formal ones anyway. What about the goal of making an AGI that can attend Stanford and get a degree, being treated the same way as an ordinary college student; or that can graduate from a purely online university? These are examples of concrete goals for AGI research. So what if everyone in the field doesn't work toward the exact same goals, instead working toward somewhat related but different goals?

The problem with this objection is, nobody ever gives a strong justification for why it's necessary for everyone in the AGI field to be working toward the same precisely defined goals, in order of the field to make progress. Yes, a shared crisply-defined goal

---

14 http://www.vetta.org/documents/UniversalIntelligence.pdf

15 http://multiverseaccordingtoben.blogspot.com/2011/03/toward-general-theory-of-feasible.html

would be a nicety (and I've put some effort into working toward one, e.g. via co-organizing the 2009 AGI Roadmap Workshop[16] at the University of Tennessee, Knoxville), but why is it posited as a necessity?

7. Objection via Consciousness

The Claim: We humans seem to be conscious; our laptops seem not to be conscious in this same sense; human consciousness seems associated with human intelligence; therefore, computers can never be intelligent like humans.

My take: Philosophers have nothing near a consensus on what consciousness is. Furthermore, philosophers have never come up with a compelling argument as to why any of us should consider our friends, wives or children conscious (as opposed to being purely solipsistic). So it seems most sensible to ignore these philosophical objections as regards AGI, until such point as a good theory of consciousness emerges. I can easily envision sitting around in a cafe' with some AGI robots debating the same old philosophical issues about who is, or is not, conscious.

My own personal philosophy of consciousness tends to be panpsychist[17] – I think everything in the universe is conscious to an extent, and each sort of system manifests this universal consciousness differently. I strongly suspect that if we build a digital mind with similar structures and dynamics to the human one, it will manifest consciousness in similar ways.

8. Objection via Free Will

The Claim: Humans have free will, we're not just deterministic automata. But a computer has no free will, it just does what its programming tells it. Human free will is closely tied with human

[16] http://web.eecs.utk.edu/~itamar/AGI_Roadmap.html
[17] http://cosmistmanifesto.blogspot.com/2009/05/blog-post.html

intelligence. Therefore, computers can never be intelligent like humans.

This notion is sometimes tied up with hypercomputing, because hypercomputing goes beyond digital computing and is hence perceived as providing a sort of free will (Selmer Bringsjord[18] has taken this perspective, for example).

My take: First of all, cognitive neuroscience has convincingly demonstrated that the human feeling of "free will" is almost entirely illusory[19]. Secondly, a large, complex computer system is not deterministic from a human perspective, in practice, so our philosophy of determinism may need some revision here (see my blog post on quantum models of classical systems[20]). And a large complex computer system becomes even less deterministic when coupled with the external world, as any realistic AGI is going to be.

9. Objection via Near-Certain Doom

The Claim: Even if we can engineer AGI, we shouldn't, because once it's completed it's going to kill us all, or enslave us, or use us for batteries, or whatever.

Some futurists, like Hugo de Garis, think AGI will almost surely eliminate humanity[21], but also think we should create AGI anyway, because it's grand and wonderful and the next step in the evolution of mind.

---

[18] http://homepages.rpi.edu/~brings/

[19] http://www.physorg.com/news186830615.html

[20] http://multiverseaccordingtoben.blogspot.com/2009/06/quantum-logic-models-of-classical.html

[21] http://www.forbes.com/2009/06/18/cosmist-terran-cyborgist-opinions-contributors-artificial-intelligence-09-hugo-de-garis.html

A variant on this idea is what I've called "The Singularity Institute's Scary Idea" – the claim, frequently put forth by various folks associated with SIAI[22], that any advanced AGI, unless it's specifically engineered to be "provably Friendly", is almost certain to kill all humans. I've pointed out in a long blog post[23] that nobody has given a clear rational argument for this position.

10. Objection via Irreducible Uncertainty

The Claim: Even if advanced AGI won't **necessarily** kill us all, we can't really eliminate the possibility that it might. No matter how carefully we engineer our AGI's ethical system and its overall dynamics, there's still the possibility that it will do something unpredictable (which could possibly be bad). After all, how can we expect to precisely predict the doings of an AGI system that's as smart as us but significantly different, or significantly more intelligent than us?

I don't have a strong refutation to this objection. I think it's substantially correct – there is irreducible uncertainty, both risk and reward that are damn hard to estimate, in a project as radical as engineering advanced AGI. I also think our best course is just to embrace this uncertainty, and try to minimize it as best we can, and try to bias developments in a positive direction. My view is that the development beyond humanity to more intelligent engineered minds is almost inevitable, and none of us would be able to stop it even if we tried. But we may – may – be able to bias this development in a more positive direction.

---

[22] http://singinst.org/

[23] http://multiverseaccordingtoben.blogspot.com/2010/10/singularity-institutes-scary-idea-and.html

9

Approaches to AGI

*This chapter, and the following one on chatbots, are updated/improved versions of material that earlier appeared in the would-be pop-sci book **The Path to Posthumanity**, that I wrote with my good friend Stephan Vladimir Bugaj. (That book was my first foray into nontechnical futurist writing, and it didn't really take off; but there was some good stuff in there.) .... This is a topic on which I have a huge amount to say, and have written a lot before and will surely write again.... The present chapter presents more of a broad-scope historical focus on the field than my other nontechnical writing has tended to do.*

As I noted in the intro above, one thing this book DOESN'T contain is a detailed overview of my own work on AGI, which I've described in a variety of technical works before, and which is currently centered on the OpenCog open source AGI platform. I'm currently (mid-2014) cooking a non-technical book called "Faster Than You Think" which will cover this ground. This chapter and the next give some additional background on the AGI field, and give some perspective on my own AGI approach, but don't dig into the details....

Creating an AGI is not a narrowly defined problem like, say, building a airplane or designing a book. It's a lot more open-ended, more like "building a flying machine" or "building a device for transmitting linguistic information" or "building a device for generating electricity." We have balloons, blimps, copters, planes, pedal-gliders, rockets, Space Shuttles, catapults... We have books, ebooks, books on tape, videos, etc. And we have solar power satellites, windmills, waterwheels, nuclear fission and fusion, garbage-to-gas, fossil fuels, fuel cells.... Each type of flying machine or electrical power generator has its own particularities – its own science and its own engineering practicalities – and the same is true with the various approaches to AGI. And just as with these other areas of endeavor, even though many different approaches may be valid, certain approaches are going to be more productive at particular points in history, based on the other technologies available and the state of scientific knowledge.

In this chapter I'll give you a bird's-eye view of some of the leading approaches to AGI that have been proposed by various scientists – both in the past and right now. I won't give equal time here to all the different AGI approaches that have been suggested or pursued, but will rush lightly over some rather large and interesting areas, in order to focus on those I consider most promising for the near term. Of course I can't know for sure my judgment is correct on this, but to get anything done in reality one has to make some concrete choices – and there are plenty of other books around, explicating the approaches I neglect.

Starting at the very high level, there are at least four broad approaches to creating advanced AGI: leveraging & networking narrow AI; emulating the brain; evolving AGI via artificial life; or the direct approach of designing and engineering AGI now based on the various sorts of knowledge available. Each of these approaches then yields a host of sub-approaches, often with dramatic differences. The direct approach is the one I'll focus on most in this book, though with occasional nods to the brain emulation approach. But first, I'll run through the other approaches, highlighting their strengths and weaknesses, and explaining in each case why I've decided it's probably not the way to go if your goal is to create a beneficial AGI as rapidly as possible.

It's worth remembering that, in spite of all the diversity of opinions about the details, there is a broad agreement among AGI researchers about some basic philosophical principles. For instance, there's basic agreement in the AGI field that mind is not intrinsically tied to any particular set of physical processes or structures. Rather, "mind" is shorthand for a certain set of patterns of organization and dynamics in systems, that are correlated with generally intelligent behavior. These patterns of organization and dynamics obviously can emerge from a human brain, but they could also be made emerge from a computer system (or a whale brain, or a space-alien brain). Unless it's created via detailed brain-simulation, a digital mind will never be exactly like a human mind—but it may still manifest many of the same higher-level structures and dynamics. The disagreements in the field regard how to figure out just what the critical "patterns of organization and dynamics" are, for generating generally intelligent behavior.

AGI Via Narrow AI?

As I've emphasized already, most of the AI programs around today are "narrow AI" programs – they carry out one particular kind of task intelligently. You could try to make an AGI by combining a bunch of juiced-up narrow AI programs inside some kind of overall framework. Many folks find this kind of approach appealing, for the obvious reasons that some pretty good narrow AI programs already exist, so if we could leverage all the work that's gone into them to build AGI, that would be a wonderful effort-saver. Plus there are strong economic incentives for the creation of narrow AI, due to the latter's ability to deliver commercial value relatively rapidly with fairly high confidence – so if all this short-term-oriented work could be leveraged to yield AGI with long-term value, then society could get long-term value without having to defer gratifications and place long-term progress above short-term gain. Wouldn't that be nice!

Unfortunately, I'm rather skeptical of this approach as a road to powerful AGI – basically because none of the narrow AI programs have the ability to generalize across domains, and it is far from obvious how combining them or extending them is going to cause this to magically emerge. Overall, I think this is an approach that seems appealing from a bird's-eye view, then seems to make less and less sense as you dig more and more deeply into it. When you look at the narrow AI components available today – say, Google, Deep Blue, credit card fraud detection systems, bio data analysis tools, theorem proving assistants, and so forth – the notion that you could combine them together to form a human-level general intelligence starts to seem almost ludicrous. Where would the common sense knowledge, and the ability to gain and improve it, come from?

The disparate AI approaches embodied in various existing narrow AI systems do not typically involve compatible notions and representations of what knowledge is, and how to reason about it. And even if one were to modify the underlying algorithms to be compatible, just chaining together a bunch of unrelated algorithms into a processing pipeline or interconnection network does not seem likely to produce general intelligence. It seems to me that, to combine narrow AI programs together in intelligent way would require some kind of highly capable "integrator" component, to figure out how to combine the different narrow AI systems in a context-appropriate way, and fill in the gap when none of them fits the bill. But then this integrator component would essentially need to be a general intelligence unto itself – so really one would have an AGI with various narrow-AI systems as modules to reference as needed, rather than an AGI built of narrow AI systems. The question then becomes how useful a crutch would the narrow AI systems be for the central AGI component.

None of these arguments, however, are meant to imply that hooking together narrow AI algorithms into a common adaptive framework is a bad thing to work on. I'm sure that composite narrow AI systems of this sort can be built to great effect, for various applications. For example, one can imagine an amusing and useful chatbot, along the lines of the Siri "virtual assistant" chatbot acquired by Apple in 2010, combining simplistic not-so-intelligent English conversation with narrow AI algorithms for responding to particular commands: narrow AI for travel reservations, narrow AI for calendar management, narrow AI for answering factual questions, etc. However, such a virtual digital assistant would be quite different from a generally intelligent AI capable of carrying out an English conversation with human-like understanding.

AGI Via Brain Emulation?

More and more very smart, serious neuroscientists are starting to talk about the possibility of creating human-level AGI via brain emulation. I think this is a fascinating, important approach that is bound to work eventually. I've spent a bit of my own time on related pursuits – for instance, I edited the first journal issue devoted to "mind uploading" (taking specific human brains and replicating them digitally), and I've done some brain simulation work for government customers, as part of my AI consulting business. However, my current feeling is that brain emulation won't be the fastest or best approach to creating human-level AGI.

One "minor problem" with this approach is that we don't really understand how the brain works yet, because our tools for measuring the brain are still pretty crude. Even our theoretical models of what we should be measuring in the first place are still hotly debated. Computer scientists like modeling brains as "formal neural networks" in which neurons, the main brain cells, are represented as objects similar to electrical switches. Computational neuroscientists currently like "spiking neural net" models, that go one level deeper, and look at the real-time electrical signals coming out of neurons. But then a few neuroscientists emphasize the importance of phenomena not encompassed in these spiking models (e.g. dendro-dendritic synapses, spike directivity, extracellular charge diffusion). These additional phenomena (if indeed they're important, which is currently unclear) could be incorporated in more detailed computational models, but we now lack the data to construct these accurately. A few renegades believe we'd need to go further down and model the brain at the molecular or atomic level, to model the dynamics giving rise to thought with sufficient accuracy. The point is, we understand so little about how the brain works, that we don't even know what *kind* of computational model would be needed to make an AGI that emulates the brain.

Neuroscience knowledge is increasing rapidly, and it's an exciting field to be involved in, even peripherally as I have been. fMRI and other brain imaging techniques have taught us a lot about which kinds of mental activity take place in which brain regions; and our models of low-level, localized neural activity get more and more accurate. However, we still have very little idea how knowledge is represented in the brain, except a few simple cases of sensory or motor knowledge; and the dynamics of complex thoughts remains almost entirely a mystery. Eventually these puzzles will be solved, but it will involve breakthroughs in brain imaging technology, providing massively greater spatiotemporal accuracy across the whole brain than anything we know how to build now. These advances will happen – but it's a mistake to say "we can just make AGI by copying the brain" as if this somehow reduces the problem to something simple or straightforward. Radical breakthroughs in brain imaging will be needed to accomplish this. Some may feel these breakthroughs are somehow more straightforward than the work needed to engineer AGI systems, but I don't see it that way. Personally, I feel I have a reasonable understanding of how to make a human-level AGI, and I don't know how to make radically more effective brain imaging technology! Of course, there is surely another researcher somewhere who feels he has a good idea of how to make radically better brain imaging technology, but has no idea how to engineer an AGI, and I freely admit that predicting the timing of radical advances is a difficult art. What I protest against is the assumption, which some futurist pundits seem to make, that a big leap in brain imaging is somehow intrinsically more straightforward or predictable than a big leap in AGI.

It also seems likely that an AGI modeled closely on the human brain would require drastically more computational resources than a comparably generally intelligent AGI created with a cognitive architecture more suited to the computers available today. Neural wetware is efficient at doing certain kinds of low-level operations, and contemporary computer hardware is efficient at doing other kinds of low-level operations. Achieving a given cognitive function via emulating neurons on a contemporary computer, is almost certainly never going to be the most efficient way to get that function done on that same computer – the neural approach is almost surely going to take far more memory and processing power than needed.

This inefficiency issue certainly doesn't make AGI via brain emulation infeasible. Computing power gets cheaper all the time, at an impressive rate; and if we knew how to make an advanced AGI using neural modeling, we'd have a lot of motivation to figure out clever software strategies for making its implementation more efficient. And there is also the possibility of creating custom hardware for brain emulation, varying on current hardware designs in a way specifically tailored to make brain emulation more efficient. This may well be feasible, but it's hard to say much about it in detail till we're clearer on how low-level a brain simulation needs to be to give rise to cognitive phenomena appropriately. Several research groups are already making "brain chips", including Dharmendra Modha's group at IBM and Kwabena Boahen's group at Stanford, but it's currently quite unclear whether these brain chips are going to be adequate for brain emulation. They lack mechanisms supporting efficient simulation of many neural phenomena of possible importance for cognition (directed spiking, dendro-dendritic synapses, extracellular charge diffusion and more) – but it may be that these phenomena aren't so critical for AGI after all, we just don't know at this point.

A different sort of issue regarding brain emulation as an approach to AGI is the fact that once you succeed, what you get is something with a very humanlike mind – and it's not clear whether this is a bug or a feature, in the scope of things. Digital humans would be fantastically interesting for science, but would suffer many limitations compared to digital minds created with a more flexible cognitive architecture. Creating non-human AGIs customized to serve humans is one thing, ethically and aesthetically speaking: creating digital humans and tweaking them to want to serve biological humans is another thing, and could backfire in various ways tied to the human motivational and emotional structure. Also, the maxim "power corrupts and absolute power corrupts absolutely" is an observation about human psychology rather than a universal maxim applicable to all general intelligences. Once digital human minds become possible, it's fairly likely someone will try to endow one with great pragmatic power – a scenario that's been explored repeatedly in dystopic science fiction.

On the other hand, futurists Anna Salamon and Carl Shulman have argued that human brain emulation is the safest route to creating advanced AGI, because human cognitive architecture is something we understand at least vaguely (and will likely understand better once we get to the stage of being able to construct digital humans). In their view, creating an AGI via means other than brain emulation is sort of like sampling a mind-architecture randomly from the space of mind-architectures (albeit with a certain complex set of statistical biases), and the odds of picking one that will be nice to humans is probably pretty small. I don't share this perspective, because I believe one can architect AGI systems with more rational, predictable and goal-focused cognitive architectures than humans have. Of course we can't have absolute confidence what such an AGI system will do, but there's also no absolute confidence about the behavior of hypothetical digital humans. Obviously, the level of uncertainty about such matters is sufficient that rational, deep-thinking people can be expected to have some level of disagreement.

One thing that seems very likely is: If we did create a digital human before creating some other kind of AGI architecture, then we could learn an awful lot about brain, mind and AI by experimenting with that digital human. Before too long, via experimenting with various changes to digital human minds and seeing their effects, we would gain a vastly improved scientific understanding of how human-like minds work, and also to some extent about minds in general. So creating a digital human would likely lead, before too long, to the creation of not-so-human-like AGI minds. And conversely, a sufficiently intelligent engineered AGI system, if well disposed toward helping humans with their endeavors, would probably be very helpful to biologists in their quest to scan brains and create digital humans.

Finally, it's important to distinguish brain emulation as an approach to AGI, from "neural networks" as an approach to AI or AGI. What are called "neural networks" in computer science and mathematics are formal constructions, loosely inspired by the network of neurons in the brain, but by no means constituting realistic neural models. Not many computer scientists or mathematicians get confused about this, but journalists often do.

When I was a math professor, I once had a colleague, freshly immigrated to the US from mainland China, who had done his mathematics PhD thesis on a certain kind of neural networks. He was amazed when I pointed out to him that the word "neural" referred to certain cells in the brain. His thesis had been a good one, proving novel and powerful results about certain aspects of certain mathematical neural nets – but actually, at no point in his research would it have been useful to him to know the biological analogy to the mathematical objects he was working with. And these particular "neural nets" that he studied were abstracted far enough from the original biological basis of neural nets, that it's arguable they didn't even deserve the name anymore – though they were certainly "broadly neural-network-like systems."

The popular confusion about the status of formal neural nets as brain models rose to the fore in 2009 when Dharmendra Modha's team from IBM made headlines by reporting a "computer simulation of a cat brain." Actually, what they did was to run a formal neural net model with roughly the same number of neurons as a cat brain. It wasn't a serious computational neuroscience simulation, it was a computer science style formal neural net. And furthermore, the connections between the formal neurons were decided randomly based on some statistical formulae – there was no effort to model the connection structure of a cat brain, which is what, in the hypothesis that the cat brain is well modeled as a neural network, would encode the cat's mind. So: yes, Modha's team did create a formal neural network that in some sense was the size of a cat brain. But this neural network was not the kind that models the brain in any serious way.

The same confusion popped up again in 2011 when Modha's team announced the creation of a prototype "brain chip." Similar chips had been prototyped before, e.g. by Kwabena Boahen's team at Stanford, but this naturally attracted more attention because it was IBM doing it. But the "brain chip", inasmuch as can be limned from the news reports, constitutes a hardware implementation of a relatively simple formal neural net model. This may be very interesting and useful, but it's not nearly as brain-ish as most people reading the press releases are probably led to believe.

On the other hand, in 2008 when Gerald Edelman and Eugene Izhikevich performed a large scale computer simulation of large portions of the human brain, using much more biologically realistic models of neurons and making more of an effort to emulate realistic connection statistics and brain architecture, the media hardly noticed. Their simulation demonstrated large-scale activity qualitatively resembling that in the human brain. It certainly wasn't an AGI system, but it was outstanding computational neuroscience. But IBM wasn't involved, and Edelman and Izhikevich were more conservative in phrasing their accomplishments publicly.

AGI Via Artificial Evolution

Human and animal intelligence were produced via evolution – so why not try to evolve AGIs? Evolutionary algorithms – software processes emulating the logic of evolution by natural selection – have already proved effective at finding solutions to various simpler problems. And "artificial life" software programs have showed the artificial evolution of simple digital organisms in simple simulated environments – such as artificial video-game-like bugs, or simulated DNA that emerges from a simulated prebiotic soup and reproduces itself, etc. Why not take this one step further and just run some kind of computer-simulated ecosystem, and let generally intelligent simulated organisms emerge therein, selected via evolution to achieve their goals in the simulated environment?

There are a couple of reasons why this probably isn't the best approach to AGI right now – though it's certainly an interesting research direction. First of all, we don't understand evolution terribly well yet. Secondly, the amount of computing power required to accurately simulate a rich, diverse ecosystem from which intelligence could plausibly emerge, would clearly be vastly greater than the amount required to accurately simulate a single intelligent system.

The natural counterargument to these complaints would be that maybe we don't need to emulate biological evolution exactly, but just to implement some broadly similar evolutionary process in a simulation... And maybe we can tailor this simulation to foster the evolution of general intelligence, so that the computational requirements won't be so severe as in a simulated environment not thus tailored. This counterargument is reasonable enough on the face of it, but my own experience playing with "artificial life" type simulations – and following the literature in the field -- has made me very aware of the difficulties faced in trying to design a really effective simulated evolutionary process. So far, every artificial life simulation ever created hits a fairly low ceiling of emergent complexity – none of them, so far, has led to a series of surprising emergent phenomena (new digital organisms, new behavioral phenomena) going significantly beyond what the programmers expected when they created the simulation.

Little simulated bugs, put in an artificial environment where they need to compete and cooperate for food, will run around and compete and cooperate for food and demonstrate funky behavior patterns, and it's all a lot of fun – but they don't evolve dramatic new structures or behaviors or get dramatically smarter and smarter or give rise to new, unforeseen kinds of organisms. Alife pioneer Tom Ray got DNA-like replication behavior to emerge from an impressively simple Alife substrate, but his followup project, aimed at demonstrating a similar emergence of multicellular digital life from single-celled digital life, failed to yield the hoped-for results.

My own impression, from working on Alife for a while in the 1990s, was that making really rich and fertile Alife would require a deeper artificial infrastructure than is typically used in the field. In the typical Alife application, one has artificial organisms encoded by artificial genomes, which then are adapted via evolution based on fitness in the artificial world. But the mapping from genome to organism – from genotype to phenotype, in the biology lingo – is extremely simplified. In real organisms this mapping involves all sorts of complex mysterious processes like protein folding (which we still don't understand well), spatiotemporal pattern formation, nonlinear-dynamical self-organization and all manner of signaling. Which aspects of the genotype-phenotype mapping are important for the creative richness of the evolutionary process as it's happened on Earth?

The biological genotype-phenotype mapping involves a heck of a lot of chemistry, and this seems to be important in terms of evolution's ability to come up with wacky new things that were unpredictable based on the organisms that came before. All these wacky new things have to operate consistently with the laws of chemistry and physics, and are guided by these laws in their particular manifestations on the surface of the Earth. This led some Alife researchers to work on "artificial chemistry", and try to ground their artificial biology in artificial chemistry. But none of the artificial chemistries seemed to demonstrate the creative flexible generativity of real chemistry. Which aspects of real-world chemistry are important for the emergence of biology that's capable of leading to the evolution of general intelligence in environments resembling the one on the surface of the Earth?

And chemistry, of course, gets its power and particularities from physics... So the question may even become, what aspects of physics are necessary to give rise to a chemistry capable of giving rise to a biology capable of fostering the evolution of general intelligence given a reasonable amount of time and space and materials? I very much doubt one needs to simulate physics down to the subatomic level in order to make an artificial life simulation giving rise to the same flexibility and generativity as real biology. My intuition is that some sort of fairly abstracted, well-designed artificial chemistry should do the trick. However, my main point here is that some fairly deep open research problems are involved here.

As in the case of brain emulation, my point isn't to argue against Alife as an area of research – I think it's a fascinating field and that people should be working on it! I wish all these areas of research were funded a lot more heavily, as opposed to many of the fairly pointless things our society spends its time and energy on. I'm quite confident AGI via brain emulation *will* work, though my best guess is that it will arise *after* AGI already exists via other means, because radical advances in brain imaging may not come that fast. I suspect that AGI via Alife will work eventually, though probably well after both engineering and brain emulation have succeeded, since the amount of computing power required to create a good-sized artificial ecosystem will be immense. I doubt that "AGI via networked narrow AI" will ever succeed fully, but I think the attempt might lead to some rather interesting software and hardware applications.

Good Old-Fashioned AI

What I'm calling "symbolic AI" is – in its simplest, classic form, at least – sometimes referred to as Good Old Fashioned AI (GOFAI). GOFAI systems are based on rules that are much like mathematical axioms or grammar rules. But instead of combining algebraic variables or parts of speech, these rules combine concepts. Such a system tries to solve the problems it's confronted with by figuring out how to combine its rules to yield appropriate conclusions (conclusions representing the answer to a question it was asked, or representing the series of actions that must be taken to achieve the goal it's been asked to achieve, etc.).

One formerly popular kind of symbolic AI system, an "expert system", used rules that were explicitly encoded by human experts to represent their specialized knowledge about some subject. To give a concrete flavor of this kind of AI, here is a simple English version of a typical rule used by the classic medical expert system MYCIN:

IF

1) the stain of the organism is gramneg and
2) the morphology of the organism is rod and
3) the aerobicity of the organism is aerobic

THEN

there is strongly suggestive evidence (0.8) that the class of the organism is enterobacteriaceae

The rule isn't actually expressed in English inside the expert system, it's expressed in a mathematical language; but it can print out the English version for human consumption.

And the following shows an example interaction between a human and a medical expert system that uses rules like this internally:

1. Patient Information
    1) Patient's name
       PT538
    2) Age
       34 years
    3) Sex
       Male

Diagnosis

6) Please enter results of pending cultures in table:
SITE CULTURE# DATE EXAMINED
CSF 1234 30.1 YES

10) Do you suspect PT538 may have an infection at a site from which you have not obtained specimens?
No

15) Has PT538 recently had objective evidence of abnormal neurologic signs (e.g. seizures, coma) documented by physician?
Yes
The CSF culture will be considered to be associated with meningitis.

Cause

17) Does PT538 have an abnormal chest x-ray?
No

18) Has PT538 been exposed to any contagious diseases recently?
No

27) Please enter CSF findings in table:
WBC %PMN Protein CSFglucose Periph-glucose
100 99 280 *5 UNKNOWN*

35) Is meningitis a hospital-acquired infection?
Yes

39) Is PT538 a burns patient?
Yes

I have found evidence (RULE545 RULE557)that treatment should cover for e-coli causing meningitis

I have found evidence (RULE545 RULE557 RULE578)that treatment should cover for pseudomaonas aeruginosa causing meningitis

I have found evidence (RULE545 RULE557 RULE578)that treatment should cover for Klebsiella pneumoniae causing meningitis

I have found evidence (RULE507)that treatment should cover for Diplococcus-pneumoniae causing meningitis

Treatment

It is important to cover for the following infection(s) and associated organism(s):

Meningitis
<item 1>e-coli
<item 2>pseudomaonas aeruginosa
<item 3>Klebsiella pneumoniae
<item 4>Diplococcus-pneumoniae

41) Does PT538 have a clinically significant allergic reaction to any antimicrobial agent?
No

43) PT538's weight in kilograms:
70

Recommendation 1
My preferred therapy is as follows:
In order to cover for items <1 2 3 4>
give the following in combination:

1) AMPICILLIN
Dose 3.5g q4h IV

2) GENTAMYCIN
Dose 119 mg q8h IV

Since high concentration of penicilllins can inactivate aminoglycosides do not mix these two antibiotics in the same IV bottle.

All this is quite impressive, and quite useful. What makes this sort of approach worrisome from an AGI perspective is that the expert system didn't really have any understanding of what it meant by terms like "dose" or "ampicillin." It knew them only as symbolic tokens, taking part in rules encoded by humans.

Expert systems can actually do very well in some areas – medical expert systems like MYCIN being one example. They have been shown to diagnose diseases better than most human physicians. Of course, such an expert system won't notice if the patient is lying or incorrect about their symptoms, but truth be told, most doctors won't either. Once the symptoms are known, the expert system can apply its rules to figure out what the cause of the symptoms is likely to be. Of course, is that the system did not learn any of the rules it uses – they were encoded by a human. But then, most doctors didn't learn the rules they use to perform diagnosis, at least not in the sense of inducing them from their experience – they just memorized them from a textbook or a teacher.

When MYCIN was created at Stanford in the early 1970s, getting diseases diagnosed by an AI program was a pretty radical idea. Nowadays, there are plenty of websites where you can click through multiple pages, choosing your symptoms from a list of options on each one, and at the end get some information on what's probably wrong with you. So the idea of getting diseases diagnosed or medicines chosen by a computer program probably wouldn't shock anyone now. In fact, we don't commonly think of such websites as AI, even though (in some cases) they're doing basically the same thing as MYCIN – we just consider them useful computer software.

An expert system is a paradigmatic narrow AI system – it carries out a very specific purpose, and in some cases it does so quite well. However, it doesn't understand the context in which it operates, and so it has severe limitations in terms of general intelligence. If a new disease is discovered, an expert system for medical diagnosis will never adapt to it, unless some human being updates its rule base.

GOFAI Grows Up: SOAR

Modern symbolic AI systems go far beyond early expert systems – perhaps the most impressive such system is SOAR, created in 1983 and developed continuously ever since, which uses hand-coded expert rules at various levels of abstraction, but wraps them in a fairly sophisticated cognitive architecture modeled on human psychology. SOAR has been used to model human psychological behavior in various laboratory experiments, and also to perform various practical tasks. Perhaps the most impressive SOAR application was TacAir, which simulated the behavior of human fighter pilots, using hand-coded rules describing their knowledge and behavior in different situations.

TacAir-Soar was a flight simulator capable of flying any routine pattern that the US Navy, Air Force, and Marines had for any fixed-wing aircraft. It could prioritize objectives, make decisions, and follow orders on its own. For example, suppose the simulated airplane takes off, with a current mission objective of reconnaissance. Suppose that partway through the mission, an enemy fighter launches a missile at it. Rather than continuing on its path, it will internally reprioritize, shifting the task of avoiding the missile above the task of continuing to perform reconnaissance. Once the missile has been avoided, it will then check to find out where the enemy fighter is, and decide what to do: ignore it because it's out of range, or engage the enemy fighter and shoot it down. Furthermore, it could receive instructions in simple English language sentences. It didn't learn to any significant degree – what happened to it on one mission didn't help it do better on future similar missions. But it was impressively functional, carrying out complex dynamic behaviors that would have been a lot more difficult to achieve in a standard software implementation, without sophisticated software architecture like SOAR.

Overall diagram of the SOAR cognitive architecture. "Productions" are technical lingo for formal rules similar to expert rules. "Chunking" is combination of existing rules to form new ones.

I know the two of the three originators of SOAR, John Laird and Paul Rosenbloom, moderately well, and they are both serious AGI thinkers. Neither of them considers themselves terribly close to creating human-level AGI, but (to take the risk of summarizing other peoples' views in a crude and concise way) they feel they're on research paths that have reasonable odds of eventually leading to human-level AGI – and that even if they don't get there during their research careers, maybe their students' students' students' students' will. Laird is still actively developing SOAR, whereas Rosenbloom is developing a new approach to AI based on mathematical structures called factor graphs, which he has embodied in an alternate cognitive architecture called Sigma. Sigma has a lot in common with SOAR, but also has some profound differences, particularly its embracing of probabilistic knowledge representation as a central aspect.

Laird, like everyone else in the SOAR community, is well aware of SOAR's limitations – it doesn't do learning or manage uncertainty in a very sophisticated way, it can't deal with large-scale sensorimotor data, etc. However, the basic perspective underlying SOAR is that, if you get the core cognitive architecture right, then these other aspects can be inserted into it without major changes. Rosenbloom has deviated from this perspective a bit, focusing on developing factor graphs as a single mechanism for handling every sort of knowledge. To put it crudely, he figures SOAR has already largely solved the cognitive architecture problem, but to make a powerful AGI system, it will be necessary to have a stronger underlying knowledge representation framework, which can handle large amounts of uncertain knowledge in a scalable way – and he thinks factor graphs can fit the bill.

Anyway, much as I respect Laird and Rosenbloom both, I think their approaches to AGI (and especially SOAR) still have far too much GOFAI in them. In my view, beginning with hand-coded rules starts you off in the wrong direction – because knowledge learned from experience has a dramatically different character from hand-coded knowledge. Experientially learned knowledge tends to be far less crisp and well-organized than human-coded rules, it consists of complex networks of interrelated, weakly-held knowledge; and it's appropriately handled by cognitive processes very different than the ones that work on small sets of crisply defined rules. Humans can indeed handle crisp, formalistic rules like those existing in expert systems or SOAR – but they do so using methods that are evolved to be guided by other, unconscious cognitive processes acting on huge networks of highly uncertain knowledge at varying levels of abstraction.

Millions and Millions of Rules: Cyc

The ultimate monster GOFAI systems is actually one of the more recently conceived ones: Cyc – a project ongoing since 1984, founded by Doug Lenat. In good old GOFAI fashion, Cyc is focused on trying to build a program with common sense, via programming it explicitly as a massively interconnected rule set. The main thing differentiating Cyc from prior GOFAI efforts is the sheer bulk of its rule-set. The Cyc team is mainly focused on encoding millions of items of data, so that the program can know everything an eight-year-old kid knows.

"Cyc" was originally short for "encyclopedia," but they found that the knowledge they needed was quite different from that found in encyclopedias. It turns out encyclopedic knowledge is too abstracted and requires too much grounding to serve as *foundational* for intelligence development, something which should have been more obvious form the start. Rather, they are now focused on everyday knowledge you could get by asking a small child: a combination of dictionary-like simple word definitions and contextually-embedded situational knowledge. Each common-sense concept in Cyc gets an English-language definition as well as a mathematical definition, which tries to paraphrase the English definition. For example, the Cyc-English definition of "skin" goes like this:

> *"A (piece of) skin serves as outer protective and tactile sensory covering for (part of) an animal's body. This is the collection of all pieces of skin. Some examples include "The Golden Fleece" (representing an entire skin of an animal),*

and Yul Brenner's scalp (representing a small portion of his skin)."

The Cyc-English definition of happiness is:

> *"The enjoyment of pleasurable satisfaction that goes with well-being, security, effective accomplishments or satisfied wishes. As with all 'Feeling Attribute Types,' this is a collection—the set of all possible amounts of happiness one can feel. One instance of Happiness is 'extremely happy'; another is 'just a little bit happy.'"*

Cyc is based on getting humans to bootstrap the conceptual groundings normally gained through experiential learning by encoding their own experiences as a foundation for intelligence – in other words, to tell computers what symbols (words in these cases) mean.

The Cyc project is interesting, but I think it's fundamentally flawed – don't believe that the logical definitions in Cyc have all that much overlap with the kind of information contained in the mind of an eight-year-old child. We humans aren't even explicitly aware of much of the information we use to make sense of the world: not in the least because it has evolved through dynamic processes in interaction with our environment in complex ways, rather than being mere memorizations of formally presented definitions and rules.

A human's notion of happiness or skin is much bigger, more disorderly and messier than these definitions. Attempts to make our human definitions more rigorous and formal, and more compact, always lead to incomplete notions of things. These kinds of general abstract definitions may be inferred in the human mind from a whole lot of smaller-scale, practical patterns involving skin and happiness, but they're not the be-all and end-all. In dealing with most practical situations involving skin and happiness, we don't refer to this kind of abstraction at all, but we use the more specialized patterns that the general conclusions were derived from—either individually or in combination.

Additionally, our mental flexibility allows us to spontaneously derive new patterns from new observations, combined with old knowledge.

This is a fundamental problem with any system that cannot learn fundamentals from experience – there has been an interpretation, formalization, and compacting step by the bootstrappers in an attempt to make the task tractable which necessarily leads to information loss. Furthermore, it is biased by the originating individual's notions of the concepts, which are not 100% the same from person to person. A learning system, like a child, can take in the varying and sometimes opposing ideas about a concept, synthesize them, and make its own conclusions. A system that cannot do that is at the mercy of its initial conditions.

Basically, Cyc tries to divorce information from learning. However, this can't really be done, at least not anywhere near as thoroughly as the Cyc folks would like. In practical terms, a mind can only make intelligent use of information that it has figured out for itself—or else is of roughly the same *form* as the information it figures for itself. Furthermore, if it has not learned through experience, it has no experience doing so, and thus will be unable to properly adapt over time to a changing environment (and as any English speaker should understand, even the linguistic landscape is an ever-changing one).

It must have had a structure for autonomously-inferred knowledge in the first place; otherwise there is no mechanism for grounding knowledge in any realm of self-action. If the information read into an AI system from a database is too different in structure from the information the AI system has learned on its own, then its reasoning processes will have a hard time integrating the two. Integration may be possible, but it will be extremely time-consuming, and would most likely proceed in a similar manner to how we as humans learn information that's just dumped before us.

Any AI system will need to learn some of its knowledge base for itself—no Cyc-like system can contain all possible knowledge an AI system will ever need to know for interaction with any environment, as it's totally impossible to predetermine every eventuality it may encounter during its existence. Any knowledge base that isn't structured in a way that naturally matches with the structure of learned knowledge will be effectively useless, but how can the people building a database like Cyc know what knowledge representations will match with learned knowledge, if they aren't building a learning system? For a learning system a system like Cyc is, at best, a kind of basic encyclopedia that a sufficiently advanced learning system could learn to read – but which is not foundational to the system becoming intelligent in the first place.

Despite more than twenty years of development, Cyc never succeeded in emulating an eight-year-old child. Nor has anyone yet found much use for a CD-ROM full of formal, logical definitions of common-sense information. The company Cycorp is doing OK, supported largely by government research grants. Cycorp has been quite careful not to mention anything about artificial general intelligence or any long-term scientific mission underlying the firm's work. Instead, they characterize their goal as producing a useful database intended for embedding within various specialized software products. This is a worthwhile mission to be sure, but very different from the grand goal of AGI.

In fairness to Doug Lenat, I have to admit that he's a deep thinker and his computational-psychology perspective does have some depth to it— far more than is represented in the public face of Cyc. He has a reasonably solid theory of general heuristics – problem-solving rules that are abstract enough to apply to any context. His pre-Cyc programs, AM and EURISKO, applied his general heuristics theory to mathematics and science respectively. Both of these programs were moderately successful, exemplars in their field, but far from true general intelligence. Their design lacks a holistic view of the mind. In the big picture of AGI, getting the mind's heuristic problem-solving rules correct, alone, means almost nothing. Problem-solving rules gain their psychological meaning from their interaction with other parts of the mind, and their grounding in experience of and interaction with some environment. If the other parts aren't even there, the problem-solving is bound to be sterile.

EURISKO won a naval fleet design contest two years in a row (until the rules were changed to prohibit computer programs from entering), and it also received a patent for designing a three-dimensional semiconductor junction. Yet, when considered carefully, even EURISKO's triumphs appear simplistic and mechanical. Consider EURISKO's most impressive achievement, the 3-D semiconductor junction. The novelty here is that the two logic functions, "Not both A and B" and "A or B," are both performed by the same junction—the same device. One could build a 3-D computer by appropriately arranging a bunch of these junctions in a cube.

How did EURISKO make this invention? The crucial step was to apply the following general-purpose heuristic: "When you have a structure which depends on two different things, X and Y, try making X and Y the same thing." The discovery, albeit an interesting one, came directly out of the heuristic. This is a far cry from the systematic intuition of a talented human inventor, which synthesizes dozens of different heuristics in a complex, situation-appropriate way. EURISKO, in essence, solved the semiconductor design problem in the same way that Deep Blue plays chess – by recursively applying a given set of rigid rules until a solution that was optimal under the specified fitness criteria popped out. It could search a space of solutions under the given rule conditions and test them against a formal optimization criteria, but it had neither intuition about nor understanding of what it was doing.

By way of contrast, think about the Croatian inventor Nikola Tesla—probably the greatest inventor in electrical engineering history—who developed a collection of highly idiosyncratic thought processes for analyzing electricity (Citadel, 1998). These led him to a steady stream of brilliant inventions, from alternating current to radio to robotic control, but not one of his inventions can be traced to a single "rule" or "heuristic." Each stemmed from far more subtle intuitive processes, such as the visualization of magnetic field lines, and the physical metaphor of electricity as a fluid. Each of Tesla's notions involved the simultaneous conception of many interdependent components.

Problem-solving in the general sense is not merely getting lucky in applying a fixed set of heuristics to a particular problem; but creatively devising not only new heuristics, but whole new categories of heuristics (we would call these "new ways of looking at the problem"); and even new problems and categories of problems into which we can decompose a particularly vexing large problem. If Tesla could only work with the existing engineering heuristics of his day, rather than creating new ones, he'd never have created so many amazing inventions.

EURISKO may have good general-purpose heuristics, but what it lacks is the ability to create its own specific-context heuristics based on everyday life experience. This is precisely because it has no everyday life experience: no experience of human life and no autonomously-discovered, body-centered digital life either. It has no experience with fluids, so it will never decide that electricity is like a fluid. It has never played with blocks or repaired a bicycle or prepared an elaborate meal, nor has it experienced anything analogous in its digital realm. Thus, it has no experience with building complex structures out of multiple interlocking parts, and it will never understand what this involves.

EURISKO pushes the envelope of rule-based AI: it is just about as flexible as a rule-based program can ever get, but it is not flexible enough. In order to get programs capable of context-dependent learning, it seems necessary to write programs which self-organize—if not exactly as the brain does, then at least as drastically as the brain does. Hand-encoded knowledge can potentially be useful in this process of self-organization, but only if it's encoded in a way that matches up naturally with *learned* knowledge, so that the appropriate synergies can emerge. This means that one can't think about the mind by thinking only about rules; one also has to think about learning and experience, even if what's doing is creating rules to be fed into an AGI system.

The Modern, Direct Approach to AGI

The approach I advocate, for the goal of creating beneficial AGI as rapidly as possible, is to simply design and engineer a thinking machine. Don't wait for better brain imaging technology or a good theory of the particulars of the human brain; don't wait to understand the mysteries of evolutionary biochemistry or accumulate the compute power needed to simulate an ecosystem; don't mess around with networking narrow AI software that was never designed to handle common sense or broad-minded generalization; and for sure don't work on narrow AI with the illusion that it's someday going to miraculously start constituting progress toward AGI, which the last 50 years of AI have taught us is a quite different problem. Just move directly toward the goal of advanced AGI, with general intelligence at the human level and then beyond. Take our best understanding of how the mind works, and embody it in a software design; then implement it, test it, and teach it and let it loose to experience the world.

There's just one small catch with this marvelously direct approach, which is that "our best understanding of how the mind works" is a bit of a mess. There is no rigorous scientific theory of cognition, and so various researchers hold to various part-rigorous, part-intuitive theories, and pursue various "direct engineering" style AGI approaches based on these. I've been undertaking considerable effort to get multiple AGI researchers to agree on a single medium-granularity model of how the mind works, and then formulate their respective AGI approaches as different ways of filling in the details of this model. I think I may be able to corral at least a reasonable plurality of AGI researchers into this sort of conceptual agreement, but even in the most optimistic case, it will take a few more years of community-building and collaboration within the AGI community to get there.

Any categorization of the broad spectrum of direct AGI approaches will necessarily be incomplete and limiting – there will be cases that span multiple categories, and cases that don't seem to fit into any of the categories, etc.[24] But nonetheless the human mind likes to categorize, so here goes! The first dichotomy I would pose is between experiential learning based and encoded knowledge based approaches.

Experiential Learning Based: In this kind of AGI approach, the focus is on automated pattern discovery, wherein the system "grows" its own intelligence via analyzing the data collected from its sensory inputs. In this approach high level structures and skills (such as facility for language, planning etc.) are derived from incremental accumulation of regularities in sensory inputs by inductive inference. There will always be some structure built into the system, to guide what kind of learning is done, but this structure doesn't constitute specific knowledge content about the world, it's more high-level and implicit knowledge in the form of biases regarding how to learn about the world.

Encoded Knowledge Based: In this kind of AGI approach, the system is started with hand-coded knowledge, rather than primarily or exclusively learning via observation or experience. For instance, the facts that people and cats are animals might be written in some formal language (e.g. cat isa animal, human isa

---

[24] A common way to categorize AI or AGI approaches is to talk about "symbolic" (e.g. logic-based) versus "subsymbolic" (e.g. neural net approaches) – but, though this dichotomy is historically important in the AGI field, I think it's actually somewhat confusing. It works better as a classification of practical AI systems built so far, than as a classification of the larger designs and ideas underlying these systems. So I'll take a somewhat different tack here (though coming to the symbolic vs. subsymbolic distinction along the way).

animal) and then fed into the AI system from a file or via an interactive interface. Knowledge may also be fed into the system via natural language, using language parsers with specially contrived semantic analysis modules intended to enable the semantically unambiguous interpretation of simple cases of English sentences.

Hybrid: Of course, it's also possible to build AGI systems that derive knowledge both experientially and via hand-coded rules. Indeed, nearly everyone building systems relying on hand-coded knowledge rules intends to eventually have their system handle and learn from experiential data. And my own OpenCog approach, though fundamentally experiential learning based, enables loading in of knowledge rules as an option. However, in practice, nearly all in-principle hybrid AGI systems have a primary orientation toward either experiential learning or hand-coded rules. OpenCog's primary orientation is experiential learning based, as is seen from the fact that the system can meaningfully operate without hand-coded rules, but now without experiential learning. On the other hand, many of the "Good Old Fashioned AI" systems from the 1970s and 80s (e.g. Cyc and Soar, which I'll discuss in the next chapter) are clearly primarily oriented toward hand-coded rules: They can't be run without them, and in practice tailoring them for a new application is mainly a matter of adding new rules to the rule-base.

I don't think the hand-coding of knowledge a good approach to AGI, but for a long while it was the dominant approach in the AI community, so it's worth understanding because of its historical value and the influence it's had on the field. For now, though, let's focus on the experiential learning approach (including hybrid approaches that are primarily experiential learning focused). I think this sort of approach is much more promising. Experiential learning based approaches can be categorized via the amount of biasing that is provided to the learning.

Minimal Biasing: There are AGI systems that try to do experiential learning with as little a priori biasing as possible. Of course, any real-world system will necessarily have some bias, but one can try to minimize this, for example by not intentionally building in any biases.

The field of reinforcement learning based AI tends to have a philosophy of minimal biasing. For instance, at the AGI-11 conference there was an interesting debate between Richard Sutton, generally considered the founder of reinforcement learning based AI, and Itamar Arel, an AGI researcher who is also a big fan of reinforcement learning, but who advocates a hierarchically structured pattern recognition architecture (which I'll discuss a little later). Itamar's AGI architecture consists of a collection of pattern recognizers arranged into hierarchies, intended to recognize patterns in the system's experience, but biased via its very architecture to be extra-good at recognizing hierarchical patterns. On the other hand, Sutton's reinforcement learning approach doesn't build in any hierarchical structure, or any other kind of explicit structure. Sutton's view was: If a bias to recognize hierarchical patterns is important, then the system should learn that bias via experience, via studying the patterns in the data it obtains from the world. Itamar's view was: Yes, that sort of learning is possible in principle, but it would take a very long time and a very large amount of data, and after all the human brain has a built-in hierarchical structure in many regions, e.g. the visual and auditory cortex.

The point about the brain's built-in hierarchical and other structures is interesting, because of the brain's tremendous plasticity. Sometimes, if a part of the brain is damaged or removed, another part of the brain will take its place, adopting the structures previously possessed by the missing part. So it's not just that the brain has certain pre-wired structures; it has a pre-wired (i.e. evolved) propensity to *grow* certain kinds of structures when the conditions are appropriate.

AI researcher Ben Kuipers did some interesting experiments in which he got an AI system to learn that it lives in a 3 dimensional world. Rather than preprogramming the system to interpret its input data as 3 dimensional, he fed the input data to the system in a less structured way, and forced the system to infer the dimensionality of space from the data. Interesting and impressive – though not shocking, as it stands to reason data from N dimensions is going to look more comprehensible and less confusing if interpreted as coming from N dimensions rather than from a space with some other dimensionality.

Significant Biasing: While a minimal-biasing approach is elegant and interesting, many of us working on experiential-learning-based AGI systems are convinced that the most practical route is to build a considerable amount of structure into our AGI systems, so as to provide them with biases to learn particular sorts of things in particular sorts of contexts. The catch, though, is that there's no such wide agreement on what sort of biasing is appropriate.

Ultimately "biasing" is just another way of referring to "cognitive architecture" – if you have an experiential learning system, then its internal cognitive architecture is precisely what biases it to learn in certain ways, and to learn some things more easily and readily than others. However, if one has a system that is capable of revising its cognitive architecture based on its experiential learning, then the cognitive architecture is only an *initial* biasing, and could ultimately end up getting totally replaced based on what the system learns.

If you want to make a categorization of AGI approaches more fine-grained than this, things become even trickier, and there are many different directions to go in. Two ways to subdivide the AGI approaches in the "significant biasing" category are: unified versus heterogenous, and explicit versus emergent. These two dichotomies are independent of each other (either unifies or heterogeneous can be paired with either explicit or emergent) and also tend to be fuzzy in the cases of most AGI systems. But the underlying ideas are worth understanding.

Unified Versus Integrative

Unified: In some AGI approaches there is a single unified approach and architecture to the biasing-structure that is built in. One example is Itamar Arel's system mentioned above, which like Jeff Hawkins' Numenta and several other approaches involves a uniform hierarchy of pattern recognition units. Another sort of example would be a purely logic-based AGI system, in which a logical theorem-prover was used to do everything; or a pure evolutionary learning approach or a uniform neural net architecture.

The unified approach has the benefit of elegance and simplicity, and it seems to appeal to many computer scientists, I think because in computer science one is always seeking a single, simple, elegant algorithm to solve the problem at hand. The mathematician in me craves an elegant, unified approach to biasing learning as well. Maybe someone will discover one someday.

However, whenever I think about this topic, I keep coming back to the incredibly messy and heterogeneous nature of the human brain. The brain is not an elegant unified architecture – each region seems to organize and process information in its own peculiar way, via complex communications and collaborations with other regions. Each region leverages the multiple types of neurons and neurotransmitters in different ways. It's a beautiful complex mess, in a way, but it's still a complex mess, apparently totally lacking the elegance and simplicity of a well-honed computer science algorithm.

To take a very simple example, I've done some work recently building computational models of the brain's representation of physical space, e.g. the 2D layout of a town or a wooded area. One part of the brain, the hippocampus, represents a "third person" top down map of space, with neurons carrying out specialized roles in this regard, such as "grid cells" that respond to corners of a spatial coordinate lattice. Another part of the brain, within the parietal cortex, represents a "first person" map of space – actually both a face-centered map and an eye-centered map. The hippocampus and the parietal cortex work together closely to keep these maps well-coordinated with each other. The hippocampus carries out certain kinds of pattern-recognition on the map it stores, the parietal cortex carries out certain kinds of pattern-recognition on the maps it stores, and then the cell assemblies binding these regions integrate these various patterns (in ways that are not at all well understood). This is a relatively simple sort of cognitive activity, and even so we're just now barely beginning to understand how the brain it – but one thing that is clear from what we do understand, is that the brain does it using a variety of fairly specialized neural architectures. Regarding spatial pattern recognition, the hippocampus performs one kind of learning-biasing, the parietal cortex performs another, and the two work together synergetically via complex pathways of coordination. Actually, we do this particular thing much more simply and elegantly in OpenCog: We just have a single third-person-view map of the world the system knows, and represent first-person views as restricted views of that primary map.

So when I see an elegant hierarchical pattern recognition architecture like Itamar Arel's or Jeff Hawkins', my first thought is: OK, that may be a nice qualitative model of visual and auditory cortex and some parts of cognitive cortex ... But what about the olfactory (smell) and somatosensory (touch and kinesthesia) cortex, which are not so markedly hierarchically structured? What about the rest of cognitive cortex, which has as many tangled-up "combinatory" connections as hierarchical connections, because it originated evolutionarily from the reptile's olfactory bulb which is dominated by combinatory connections, rather than from a largely hierarchical subsystem like visual cortex? What about the hippocampus, what about the thalamus, and the many cross-connections between these non-hierarchical systems and the cortex?

Of course, neither Arel nor Hawkins is trying to model the brain, they're trying to build AGI systems – and it's certainly rational for them to argue that one particular architecture, which part of the brain uses, actually has the capability to do everything the brain does, even though the brain doesn't use it that widely. But my own intuition says that, even if one isn't trying to emulate the brain, there's a valuable lesson to be learned from the wild *heterogeneity* of the brain. I suspect the brain's heterogeneity is a well-justified response to the task of displaying general intelligence for the tasks humans must generally face, in the environments for which humans evolved, given the limited resources the human body possesses.

Integrative: The alternative to the unified approach is the modular or integrative approach, in which various different biasing-structures ("modules") are linked together to form a unified architecture. The category of modular systems may be further subdivided into *loosely-interconnected* (black box) versus *tightly-interconnected* (white box).

In a loosely interconnected system, the different modules are "black boxes" to each other, i.e. they can't see what each other are doing. So in a system like this, a vision processing module might pass information regarding what it sees to a semantic reasoning module, and the latter might pass the former information regarding what it should expect to see – but the communication is restricted to the exchange of packets of information like this. The two modules don't have awareness of each others' internal states, and don't try to guide each others' processing.

In a tightly interconnected system, the different modules are "white boxes" that can see significant aspects of each others' internal dynamics, with the result that they can try to nudge each others' learning in the right direction. So for instance, the cognition model could do more than just tell the vision module "expect to see a face in the dark in front of you" – it could actually help the vision module adjust some of its internal parameters to better recognize the face.

I feel strongly that tight interconnection is key to general intelligence. In my view, for powerful experiential learning to occur in real-world environments under realistic resource constraints, very substantial biasing of learning is needed, which means you need a fairly substantial cognitive architecture. Since our environments and goals have so many diverse aspects, a unified algorithm is going to be difficult to come by, and an integrative approach is probably more realistic. But having multiple modules that perceive each other as black boxes isn't likely to work, because there's no good defense against one of the modules getting stuck and having stupid ideas and failing to learn properly – you need the modules to be able to see something of each others' inner workings so they can better help each other out. The modules need to be more like close collaborators doing agile software development in a startup, than like paranoid bureaucrats passing each other formalized information, keeping each other at arm's length and on a "need to know basis." Most of the work I did when designing the OpenCog system, went into ensuring the various learning modules would work well together – according to a particular approach I call "cognitive synergy."

A large question that emerges in the context of the modular approach is: Where do you get the breakdown into modules from? Mathematically and conceptually, there are many, many ways to break down a learning system into different modules. If you don't use the brain as guidance, where do you turn? There's no systematic, universal theory of real-world general intelligence to turn to, so what do you do?

A fair number of AGI researchers, including myself, have turned to **cognitive science** – an interdisciplinary field at the intersection of psychology, computer science, linguistics, philosophy and neuroscience (with other disciplines sometimes thrown into the mix as well)! I helped found two university cognitive science programs, early in my career: one at the University of Waikato in Hamilton, New Zealand, and one at the University of Western Australia in Perth. In each case I found the cross-disciplinary collaboration extremely scintillating – and in fact, it was in the context of the cognitive science group at the University of Western Australia that I came up with the first AGI designs vaguely similar to OpenCog. (My AGI designs before that were more in the "simple and elegant" vein, based on novel self-improving learning algorithms that I implemented in the beautiful and then-very-inefficient programming language Haskell, with very little practical success.)

Cognitive science is nowhere near having a full understanding of the human mind, at this point. However, one thing it *has* achieved is a reasonable understanding of how the human mind can be meaningfully broken down into modules. The modules are not always all that distinct, and they are definitely tightly interconnected. But over the last few decades of cognitive science research, something like a majority understanding of the "high level boxes and lines diagram" of the human mind has emerge." This is not enough to guide the construction of an AGI system in any detail. However, it's enough to guide the *modular structure* of an AGI system. One is then left with the "itsy bitsy problem" of putting appropriate representational and dynamical learning mechanisms inside the modules, taking cognitive synergy into account at every step.

Emergent Versus Explicit

Overlapping with the unified/modular dichotomy is the emergent/explicit dichotomy, which has played a large and contentious role in the history of AI. This is a topic on which misunderstanding is rampant, even among highly knowledgeable professional AI and AGI researchers.

(Common terminologies are symbolic vs. subsymbolic, or symbolic vs. connectionist – but these are all terms that have been given multiple meanings by various well-known researchers. To avoid compounding the confusion that already exists regarding these terms[25], I've chosen to go with the eccentric "emergent vs. explicit" instead!)

The basic idea of the distinction is that, in an explicit system, the system's designer can give detailed answers to questions like "How does your system represent its knowledge about cats?", or "What steps does your system take to figure out how to navigate through a crowded room to get to its destination?" The designer could give a general answer about the type of representation or steps-to-improvement without looking at the system's state; and then if they looked at the system's state, they could read off the detailed answers to the questions without a huge amount of hassle.

In an implicit system, on the other hand, the only way for the system's designer to answer a question like this would be to carefully study the structures and dynamics that had emerged within the system as a consequence of its engagement with the world. In other words, the system's designer would be in almost the same position as a neuroscientist asked similar questions about a human brain – except that in the case of an AI, it's possible to gather accurate data about the internal state. So the system's designer would be in a position roughly similar to that of

---

[25] To see why the "symbolic" and "connectionist" terminology is confusing, consider: Any human-level AGI system is likely to learn symbols of some kind. If a purely implicit, self-organizing neural net system chooses to create neurons or small neural assemblies that have clear interpretations as symbols, does that make it a symbolic system, or not? And "connectionist" seems clear when you compare a GOFAI system versus a neural network, but it becomes less clear when you realize that most GOFAI systems could be reformulated as semantic networks, i.e. nodes connected to other nodes by various types of links, with rules for updating the nodes and links depending on their types. So is a semantic network implementing a GOFAI type system "connectionist" or not? ... it would certainly have a lot of connections!

a neuroscientist with really amazing brain imaging tools and a thorough knowledge of the lowest-level components of the brain.

This is obviously not a very strict dichotomy, in principle, because there can be varying degrees of difficulty in answering specific questions about how a system's internal operations correspond to its knowledge or its behaviors. But the dichotomy has assumed an outsized role in the history of AI because most of the dominant AI approaches have tended to fall very strongly on the explicit or implicit side of the dichotomy.

The GOFAI systems that dominated the AI field in the 70s and 80s tended to rely on human-encoded knowledge, and tended to store this knowledge internally in an extremely transparent way. For instance, you might feed such a system knowledge in the form "cat isa animal", and then the system would store this in an internal data structure equivalent to a node for "cat," a node for "animal," and a link labeled "isa" between them. Here the knowledge representation is extremely implicit. Similarly, the traditional AI planning algorithms follow an easy-to-understand, highly routinized step-by-step approach, so that it's pretty easy to foresee how such a planner will approach a problem like navigating to a new location.[26]

---

[26] A note to the reader who has studied a little math or AI: Many of these GOFAI systems used varieties of formal logic to carry out reasoning and learning, and because of this, the impression spread in the AI community that formal logic based systems are intrinsically tied to an explicit, human encoded knowledge based AGI approach. But actually this is not accurate. It would be quite possible to make a formal logic based AI system that operated on a pure experiential learning basis; and in this case, the representation of knowledge would most likely be implicit, as it would be difficult for anyone to figure out what combinations of perceptual logical primitives the system had learned to use to represent everyday objects and events. What's not clear is whether a purely logic-based approach could be made scalable enough to serve as a practical AGI system. I'm more optimistic about including logic in hybrid systems, and this is in fact done in OpenCog. I'm still shocked by how many smart,

On the other hand, purely "connectionist" systems such as the majority of neural net models, take a much more implicit approach – there's no easy way to see why the system has learned what it has, or how the system represents the knowledge it has acquired. Using statistical and machine tools it's often possible to perform some analysis and figure out these things, but then it becomes a research project similar to figuring out how the brain does things internally. For instance, in many cases one can apply a statistical method called "principal components analysis" to study a formal neural network, and then the method produces a series of "principal components" that approximately correspond to the main memories that the system has learned. This is very useful, but it's different than looking right into a GOFAI system knowledge base and seeing "cat isa animal."

As you've probably guessed by now, OpenCog takes a hybrid approach, with some explicit and some implicit representation. But if one side of the dichotomy had to be chosen, we'd have to choose implicit for OpenCog – according to the underlying theory, the system should be able to achieve human-level general intelligence without any explicit representations, but it couldn't do so without a lot of implicit representations.

---

professional AI and AGI researchers believe that a system containing a logic-based component is somehow intrinsically unable to carry out experiential learning! One of the ironic things about this is that OpenCog's probabilistic logic equations are real-valued functions not too dissimilar from the mathematical functions used to implement neural net models. And there are approaches to logical inference that are fully parallel and distributed just like a neural net (and more so than some neural net learning algorithms). On the math level, the logicist versus connectionist distinction doesn't amount to that much – one could argue there's a significant conceptual difference, but I think even that's been far overblown. I'm often bemused by the way our thought is directed by the somewhat arbitrary categories we adopt to guide our communication.)

If you haven't had any exposure to the AI field before, this brief run-through of various approaches to AI may have been slightly dizzying and confusing. But I'll ask you to please stick with me through the next few chapters, where I'll discuss a number of AGI approaches in detail – including many of the ones briefly alluded to above. Among other thing these chapters will give some concrete examples of what various AI systems in the various categories have done already – examples often make things a bit clearer! For the rest of this chapter, I'm going to discuss some more high-level, less nitty-gritty aspects of the quest to create powerful AGI.

Theory Versus Experimentation in AGI Development

Science and engineering generally advance hand in hand, but in different cases, one or the other may take the lead. The science of quantum physics advanced rather far before practical applications emerged. On the other hand, the Wright Brothers demonstrated manned powered flight well before we had a mathematical theory of aerodynamics solid enough to guide the particulars of aircraft design. There is a variety of opinions about whether AGI should follow the quantum physics approach, the Wright Brothers approach, or something in between.

On the "quantum physics" side, some people think that creating AGI is going to take a really elegant formal scientific theory of intelligence – and that after we have that theory, building AGI will be fairly straightforward.

Some very serious researchers are pursuing this approach – for example Marcus Hutter, a German AI researcher now based at the Australian National University, has developed an abstract theory of AGI that tells you a lot about how to create powerful AGI systems in the case where you have infinite (or at least impracticably massive) computational resources at your disposal. He's trying to "scale the theory down", and render it applicable to more realistic cases. If this approach works, then eventually AGI implementation will be guided strongly by mathematical AGI theory. His approach would fall into the "biased experiential learning" camp, and it's not clear at this stage whether a future, more practical version of his general theory would tend toward unified or modular biasing-structures. Recently he's created some nice narrow AI algorithms inspired by his general theory, which are good at solving certain sorts of pattern recognition problems.

In spite of my background as a mathematician, these days I tend to fall closer to the "Wright Brothers" camp – I think we could build an AGI now, with our current fairly thorough but only semi-rigorous conceptual understanding of intelligence. In this case, perhaps the elegant theory of general intelligence will emerge later – after we have some real AGI systems to experiment with. Certainly, formulating a scientific theory is much easier when you can carry out observations and experiments on the subject of your theory!

And I do think a theory of AGI will be science and not just mathematics, because real-world AGI is about general intelligence in the physical world; and human-level AGI is specifically about general intelligence in those environments for which humans evolved. There is a certain aspect of general intelligence that is independent of what physics your universe has, and what kind of body you have, and so forth – but I think there are also important aspects of general intelligence that depend on these things. A good theory of general intelligence would tell you how general intelligence depends on these various factors, and would then let you derive conclusions about human-level AGI design from information about the human body and the environments and tasks for which humans evolved.

In fact this is my vision of what a rigorous theory of general intelligence will look like, once we have it. I think that one will be able to create a mathematical function whose input is a description of an environment, and whose output is a description of the cognitive architecture (the biasing-structure, in the language I used above) needed to be reasonably generally intelligent in that environment giving moderately limited computational resources. I've spent some effort developing this sort of theory but it's quite incomplete at the moment... I sometimes find it hard to know how to balance my time between this sort of theory and practical system-building. I'm pretty confident the OpenCog approach can work for human-level AGI, without any massive changes or theoretical breakthroughs required. On the one hand, a theoretical breakthrough might tell me how to drastically simplify the design, which would reduce the time and cost of implementation. But on the other hand, aside from the risk that formulating a serious theory of general intelligence proves too difficult, there's also the risk that after doing a lot of theory, I just wind up with a more rigorous grounding for the sort of system I'm already building.

In spite of my particular speculations about what a rigorous AGI theory might look like, I think it's also important to realize that, at this stage, *nobody really knows what a theory of AGI would or should look like*. "Physics envy" needs to be resisted – we can't expect an equation of thought with the simplicity and power of Newton's Laws or the Schrodinger Equation. Biology and AI are not physics. Biological systems are complex and messy and one can't expect elegant unified theories like the ones seen in physics. Most probably human-level AGI systems are going to be complex and messy as well, due to the same sorts of resource requirements that pushed biological systems in this direction. Yet even complex and messy systems may obey high-level structural and dynamical principles. I expect we will discover what these are as we progress with the implementation, testing and teaching of moderately powerful AGI systems, rather than before. But I'll be pleasantly surprised if somebody comes up with a useful breakthrough in the rigorous theory of AGI in the near future (and even more pleasantly surprised if it's me, heh!).

Can Digital Computers Really Be Intelligent?

One more issue needs brief attention, before we move along. All the specific AGI approaches I've mentioned above make the implicit, unstated assumption that ***it's possible to make AGI on digital computers***. But it's worth pausing a moment to point out that this isn't proven – and there are some smart, relevantly knowledgeably people who think it's definitely not true. I don't agree with them, but their arguments are worth considering.

Some theorists argue that digital computers will never display human-level general intelligence because mind is intrinsically a quantum phenomenon – relying on peculiar properties of quantum physics, which (according to these theorists) manifest themselves in the human brain, but not in digital computers. This is actually a claim of some subtlety, given the wacky properties of quantum computing. David Deutsch proved that quantum computers can't compute anything beyond what ordinary digital computers can... ***But*** (a major but!) in some cases, quantum computers can compute things *much faster* than digital computers.[27]

A few mavericks like Stuart Hameroff and Roger Penrose have gone even further, and argued that non-computational quantum gravity phenomena are at the core of biological intelligence. That is uncowed by the theorems saying that quantum computing can't compute anything beyond what ordinary computers can, and inspired by an intuition that human cognition must involve something beyond computing, they have posited that the human brain must be some other kind of computer even stranger and more powerful than a quantum computer! Modern physics has not yet found a unified theory of quantum physics and gravitational physics, and Penrose and Hameroff have exploited this loophole to suggest that once a unified quantum gravity

---

[27] Technically, they can compute some things much faster in the average case not in the worse case.

theory is found, the brain will turn out to be a quantum gravity computer!

Now, this is very difficult to refute, since there currently is no good theory of quantum gravity, so nobody knows what a quantum gravity computer is. However, it's worth noting that there currently zero evidence of cognitively significant quantum phenomena in the brain – let alone mysterious quantum gravity phenomena!

However, a lot of things are unknown about the brain—and about quantum theory in macroscopic systems like cells; and about quantum gravity —so these points of view can't be definitively ruled out. Even if there is something to this whole line of thinking, though, there are an awful lot of unknowns to resolve before you could conclude that human-level AGI is impossible on digital computers. Maybe the brain uses weird quantum voodoo to do some of its computing, but digital computers could achieve even greater intelligence using other means. Maybe the boundary between quantum and classical systems isn't as clear as currently thought – a few physicists[28] have argued that it sometimes makes sense to model large classical systems using quantum theory – in which case it could be that digital computers are also "quantum" in some sense.

Another question is, even if aspects of the human brain's dynamics rely on quantum dynamics, how important are these to emulating the brain's functions? Stuart Hameroff, when he talks about quantum theory and intelligence, likes to point to an image of a paramecium – and point out that currently we can't even simulate a paramecium on computers, let alone a human. That's true. But we also can't simulate a log, yet we can build steel struts that are better than logs at holding up buildings. He likes to point to the incredible computational potential of the molecules in the cell wall of a neuron – but so what? The cells in my skin have

---

[28] Diederik Aerts and Harald Atmanspacher, for example.

that same computational potential, yet I really doubt the dermis inside the heel of my foot is doing a lot of highly intelligent computation.

And of course, even if quantum computing or quantum gravity computing is necessary for human-level AGI (a supposition for which there is currently zero evidence), this wouldn't rule out the project of AGI as a whole. It would just mean we have to shift to a different kind of computing infrastructure. DWave is already commercializing a limited form of quantum computing, and the quantum computing industry can be expected to explode dramatically during the next century.

Hameroff, Penrose and others tie their quantum theory of intelligence in with a theory of consciousness, basically claiming that, even if a digital computer managed to "impersonate" intelligence, it would never actually be conscious, because consciousness has to do with macroscopic quantum phenomena that can occur in brains but not digital computers. Interestingly, Hameroff is a sort of panpsychist as well, positing that everything in the universe has a bit of "proto-consciousness", but only in systems that achieve general intelligence via quantum gravity computing does this develop into full-fledged consciousness. Given the massive confusion afoot regarding both quantum gravity and consciousness, and the early-stage nature of neuroscience, cognitive science and AGI, this seems to me a terrible tangle of confusions that could be unraveled only with great effort. Personally, I'm not very worried about the quantum gravity and consciousness issue as an objection to the creation of powerful AGI! Whatever consciousness turns out to be, I have a very strong feeling that if a system displays similar behaviors and similar internal structures and dynamics to a human mind, it's going to be conscious in the same sense that a human mind is. I very much doubt it will be necessary to move beyond classical digital computers to achieve a system with this kind of behaviors, structures and dynamics – but if it does, then we'll do it!

As it happens I have spent a fair bit of time thinking about femtotech, which is a hypothetical way of building computers and other machines using elementary particles. Even if (as I suspect) quantum gravity computation doesn't play a role in the human brain, perhaps it will play a role in future AGI systems implemented using femtotechnology! I'll talk about this a bit later, in these pages.

But that brings us rather far afield, into what I like to think of as "hi-fi sci-fi." I love thinking about such topics, but I prefer to spend more of my time focused on present reality, on the relatively mundane matter of creating human-level thinking machines on digital computers.

This was originally an H+ Magazine article, with a title formulated by the magazine's editor at the time, the incomparable RU Sirius:

10

Will Bots Feel Joy?

Will machines ever really ***feel***, in the same sense that humans do?

This is – at least according to many philosophies – a separate question from whether machines can be intelligent, or whether they can *act like* they feel. The question is whether machines – if suitably constructed and programmed – can have awareness, passion, subjective experience... consciousness?

I certainly think so... But generally speaking there is no consensus among experts. It's fair to say that – even without introducing machines into the picture – consciousness is one of the most confused notions in the lexicon of modern science and philosophy.

I became acutely aware of this confusion when, in summer 2009, I organized a workshop on Machine Consciousness in Hong Kong, as part of the larger Toward a Science of Consciousness conference and Asia Consciousness Festival. The TSC conference as a whole attracted hundreds of participants, but only a couple dozen deigned to venture into the riskier domain of machine consciousness; and among these brave souls, I reckon there were *more* than a couple dozen views on the matter at hand!

First we have the materialists. Joscha Bach – a German AI researcher and entrepreneur and the author of *Principles of Synthetic Intelligence*, who will pop up a bit later in this book – summarizes their perspective elegantly: "The notion of the mind as an information processing system, capable of forming an integrated self-and-world-model, modulated by emotional configurations and driven by a finite set of motivational urges, is sufficient to remove the miracles [that some associate with consciousness]." Daniel Dennett is the best-known modern advocate of the materialist view. According to his book *Consciousness Explained*, it's patently obvious that machines can be conscious in the same sense as humans if they're constructed and programmed correctly.

Paul Fahn, an AI and robotics researcher at Samsung Electronics, presented this perspective at the MC Workshop in the context of his work on emotional robots. His core idea is that if a robot brain makes emotional decisions using a random or pseudorandom "preference oracle" similar to the one in a human brain, it will likely be emotional in roughly the same sense that humans are – and possessed of its own distinct but equally valid form of consciousness. Fahn emphasizes the need for empirical tests to measure conciousness – and Raoul Arrabales's talk at the workshop took concrete steps aimed in this direction, describing a series of criteria one can apply to an intelligent system to assess its level of consciousness.

But some, less happy with the materialist view, have referred to Dennett's book as "Consciousness Explained Away." Neuropsychologist Allan Combs has a new book in press, called *Consciousness Explained Better* – in which he reviews a host of states of consciousness, including those accessed by mystics and meditators as well as those we feel in various unusual states of mind, such as dreaming, sleeping, dying, etc. As a panpsychist: He sees consciousness as the basic material of the cosmos; he sees rocks, bugs, cows, humans and machines as differing manifestations of universal consciousness.

To a panpsychist, the question isn't whether machines can be conscious, but whether they can manifest universal consciousness in a manner similar to how humans do. And the question of whether consciousness can be empirically measured is not that critical, because there's no reason to assume the universe as a whole is understandable in terms of finite sets of finite data-items, of the sort that science works with. Setting aside mystical notions, pure mathematics points to all manner of massively infinite constructs that – if they "existed in reality" – could never be probed via scientific measurements.

The coauthor of Combs' workshop talk, creativity theorist Liane Gabora, holds the view that machines are conscious, but will never be nearly ***as*** conscious as humans. "I put my money on the argument that living things are more conscious than rocks or computers because they amplify consciousness by being self-organizing, self-mending, and autopoietic; that is, the whole emerges through interactions amongst the parts. And the human mind amplifies consciousness even further through a second level of autopoietic structure. Just as a body spontaneously repairs itself when wounded, if someone does something out of character or something unexpected happens, the mind spontaneously tries to repair its model of the world to account for this turn of events. This continuous building and rebuilding of a mental model of the world, and thus reconstituting of autopoietic structure, locally amplifies consciousness. Until computers do this, I don't think their consciousness will go much beyond that of a rock."

As a panpsychist myself, I find Liane's view sympathetic – but I'm much more optimistic than she is that complex, self-organizing autopoietic structure can be achieved in computer programs. Indeed, that is one of the goals of my own AI research project!

Then there are the quantum consciousness folks, such as Stuart Hameroff, who gave the keynote speech at the Cognitive Informatics conference in Hong Kong, the day after the MC workshop. An MD anesthesiologist, Hameroff was seduced into consciousness theory via wondering about the neurobiology by which anesthetics bring about loss of consciousness. Together with famed physicist Roger Penrose, Hameroff developed a theory that consciousness arises via quantum-mechanical effects in structures called microtubules, that make up the cell walls of brain cells.

A common joke about the Penrose-Hameroff theory is: "No one understands quantum theory, and no one understands consciousness, so the two must be equal!" But clearly the theory's intuitive appeal goes beyond this: Quantum nonlocality implies a form of interconnectedness of all parts of the cosmos, which resonates well with panpsychism.

Penrose believes that human consciousness enables problem-solving beyond what any computer can do. To bypass theorems showing this kind of capability wouldn't be provided by mere quantum computing, he proposes "quantum gravity computing," based on an as-yet unknown unified theory of quantum physics and gravitation. Most scientists view this as fascinating, highly technical sci fi.

Regarding panpsychism, Hameroff says "I disagree only slightly. I would say that what is omnipresent in the universe is proto-consciousness... Penrose and I say proto-consciousness is embedded as irreducible components of fundamental spacetime geometry, i.e. the Planck scale, which does indeed pervade the universe." He views consciousness per se as a special manifestation of proto-consciousness: "I don't think a rock necessarily has the proper makeup for the type of quantum state reduction required for consciousness."

A fascinating twist is suggested by recent work by Dirk Aeerts, Liane Gabora, Harald Atmanspacher and others, arguing that "being quantum" is more about being susceptible to multiple, fundamentally incompatible interpretations, than about specific physical dynamics. In this sense, consciousness could be quantum even if the brain doesn't display nonclassical microphysical phenomena like quantum nonlocality.

At the time of the MC workshop, my close friend and colleague Hugo de Garis was running a project at Xiamen University called the Conscious Robotics Project. However, he was among the least confident participants regarding the workshop's topic: "Explaining what consciousness is, how it evolved and what role it plays is probably neuroscience's greatest challenge. If someone were to ask me what I thought consciousness is, I would say that I don't even have the conceptual terms to even begin to provide an answer."

Australian philosopher David Chalmers – whom I got to know a bit in 2011, when we both spoke at the Australian Singularity Summit -- provided an important clarification of the consciousness issue when he introduced the notion of the "hard problem" of consciousness – namely, building the link between the physical processes and structures and behaviors associated with consciousness, and the actual experience of consciousness. He contrasts this with the "easy problems" (which are easy only in a relative sense!) such as characterizing the nature of subjective experience, and figuring out what cognitive and neural processes are associated with the experience of consciousness. The easy problems may not really be easy, but seem the sorts of things that can be solved by systematic effort. The hard problem seems a basic conceptual conundrum.

In response to this hard problem, Chalmers' own conclusion about consciousness seems to amount to a weak form of panpsychism, in which it's admitted that everything in the universe has a little bit of "proto-consciousness", which manifests as full consciousness only in certain entities. I'm not sure if this is essentially different from my own view that I call "panpsychist," in which I view everything as a little bit conscious, and certain entities as manifesting reflective, deliberative consciousness.[29]

---

[29] Another area of Chalmers' interest is the nature and prevalence of "verbal disputes" in philosophy – he's tried to formalize the notion of a

Basically he concludes that the best way to bridge the gap posed by the hard problem is to posit some sort of common substance binding together the subjective-experience and objective structural/behavioral realms.

I've thought a great deal about some of the "easy" problems of consciousness, in particular the aspects of human consciousness via which we can reflect on ourselves, and via which our consciousness analyzes itself, thus creating "consciousness of consciousness of consciousness of..." While in a sense this is an infinite reflexive process – because in this sense consciousness contains itself, and only infinite entities can contain themselves – it can be approximated by finite structures such as occur in the human brain as it's modeled by physics. But this is a part of the story I'll come back to a little later!

Another point worth emphasizing is that, among my colleagues working on OpenCog, there's a large diversity of opinions on consciousness. Panpsychists are probably in the minority. It seems the practical work of engineering a mind is largely independent of philosophical issues regarding consciousness – within limits, at any rate. If you believe that consciousness is inextricably tied up with quantum phenomena, and also that consciousness and intelligence are intertwined, then you're obviously not going to like AI approaches focused on digital computation!

Regarding the hard problem – I think one very real possibility is that we might create human-level, human-like AI systems before we puzzle out the mysteries of consciousness. These AIs might puzzle over their own consciousness, much as we do over ours. Perhaps at the 2019 or 2029 Machine Consciousness workshop, AIs will sit alongside humans, collectively debating the nature of

---

verbal dispute, and has argued that many disagreements in modern philosophy are largely verbal rather than substantive disputes. It may be that the difference between Chalmers' panprotopsychism and my panpsychism is merely or largely a verbal dispute!

awareness. One envisions a robot consciousness researcher standing at the podium, sternly presenting his lecture entitled: "Can Meat Feel Joy?"

11

Deep Blue Reflections

I wrote the first version of this essay in 1997, a few days after Deep Blue's classic chess victory. It was written for the Simon's Rock College alumni email list, in response to some other alumni asking for my thoughts on Deep Blue's achievement, since I was the best-known (and maybe the only?) AI researcher on the list.... I tweaked the essay slightly at some later date....

If you're both ancient and geeky like me, you may remember May 11, 1997, the day when a computer program defeated the (human) world chess champion -- an event that led many people to think that computers were already on the verge of rivaling human intelligence. More precisely, that was the day when, for the first time ever, a computer defeated the world chess champion in a standard five-game match. Deep Blue, a computer chess system developed by Carnegie Mellon University and IBM, split the first two games with Garry Kasparov. The second two were draws, and the final game went to Deep Blue. Kasparov was a sore loser. Deep Blue remained dispassionate – it hadn't been programmed with emotions.

Admittedly, this was only one match, but the tournament was not a fluke. Previous versions of Deep Blue were already able to consistently beat all but the greatest chess grandmasters prior to the Kasparov match. And computer hardware has gotten far better since then, enabling the same basic algorithms used in Deep Blue to play even more powerfully. The capability of the human brain, on the other hand, has remained essentially constant since 1997.

Although Deep Blue is not a very generally-intelligent entity according to our definition, there's something to be learned from a study of its accomplishments and the mechanisms underlying them. Deep Blue follows the same rules as human chess players, but it doesn't think at all like humans. Human chess players use geometric intuition and a sense of the flow of a game. Deep Blue calculates every single possibility; then calculates all the possible consequences of each, gives them weights based on the probability and desirability of such a next move from its "perspective," and finally picks the next move by culling-out that which produces the best set of the next N moves for some number N.

Computer programmers call this recursive logic. It applies these same fixed rules over and over and over again, constantly referring back to the results it just obtained, and figuring out how well it is doing. Because it can weight play options based on prior games, it is a "learning" system by the ML definition, but not a generally intelligent one.

Human beings might use recursive logic to play a very simple game, such as tic-tac-toe, which has very few choices. But even in tic-tac-toe, our opponents would probably object to our taking the time to calculate out the potential consequences of every possible move. Our minds are much too slow to play chess that way, nor would the game be any fun if we could. Computers, by contrast, are much, much quicker at this kind of task and do not get bored, so recursive logic can work well for them.

Of course, every chess player extrapolates, thinking: "What is the other player likely to do next? And if he does that, what am I going to do? And if I do that, what is he going to do?" But in humans, this kind of reasoning is augmented by all sorts of other thought processes, many subconscious, which constitute what we call "intuition" about a domain. What intuition really is is the application of both knowledge so deeply learned that it becomes subconsciously reflexive to apply it, and a complex type of "educated guesswork" regarding the situation taken as a whole.

For Deep Blue, this kind of extrapolation is pretty much the whole story—and it does it very, very well. Computers can extrapolate faster and further into the future than any human. The 1997 version of Deep Blue could evaluate about two hundred million different board positions every second. This figure can easily be increased for the cost of additional circuitry, but it doesn't make Deep Blue truly intelligent. And this difference in "thinking" is not trivial – it is fundamental to why Deep Blue is in no danger of suddenly becoming Skynet.

One way to understand the difference between Deep Blue and human players is to think about strategy versus tactics. There is a certain kind of creative long-range strategy that human chess grandmasters have, but Deep Blue lacks. Deep Blue makes up for this lack by elevating tactics to such a high level that it assumes the role of strategy. Deep Blue is not entirely without strategy: it carries out its superhuman tactical evaluation within the context of a collection of pre-programmed strategies, and it is capable of switching between one strategy and another in response to events. But it does not "think" strategically, it only "thinks" tactically. Deep Blue doesn't make long-range plans involving an understanding of the overall structure of the board as a dynamical system which changes over the course of the whole game, or the emotional states of its opponents. It does not innovate new plays based on an insightful, creative analysis of chess theory, but only by accident if its brute force methods stumble upon a more optimal solution. If it could do such things, it would doubtless play even better. Even without strategic creativity, it plays well enough to beat the best humans, but only because it turns out that chess is a game which succumbs well to a recursive logic approach.

The defeat of Kasparov by Deep Blue was symbolic because chess is the most mentally challenging game commonly played in the Western world. Computers became better than humans at checkers and many other games quite some time ago. However, there is at least one popular game that still stumps the best computers—the Asian game called Go (or Weixi, in Chinese). At the present time, in spite of a substantial research effort, no existing computer program can play better than the advanced beginner level at Go.

The rules of Go are very simple, compared to chess. Play is on a 19x19 grid, and stones (pieces) are placed on the intersections of the grid, called points. The first player plays black stones, the opponent white ones; and stones are added to the board one-by-one, players alternating. Stones are not removed once placed, but stones and groups of stones may be captured. A player calls "Atari" when a capture can occur on their next move, to warn the opponent. The game ends when it is no longer possible to make a reasonable move. The winner is determined by the amount of territory surrounded, less the number of their stones captured.

The trouble with Go from the computational perspective is that from any given board position, there are hundreds of plausible next moves rather than dozens, as in chess. Extrapolation in Go will not get you as far as it does in chess. It would seem that, if computers are to conquer Go, they're going to have to do it either with a more generally intelligent approach, or use some more clever special-case technique than the one employed for chess. Go is too visual, too two-dimensional, to succumb to purely combinatorial, non-visual techniques. A world-champion Go program would have to be intelligent at general two-dimensional vision processing as well. Since Go, like chess, is ultimately an extremely limited problem domain, a special-purpose, unintelligent program may possibly master it eventually by employing clever domain partitioning and parallelization techniques in a brute force recursive logic approach, probably coupled with hardware massively more powerful than current systems.

However, the fact that it hasn't been mastered yet just goes to show how far from intelligence computers really are right now—there are even very narrow domains with which they can't really cope at this time. Human Go masters may have little or no other things that they are true masters of, but they are not fundamentally restricted to a single domain like Deep Blue is – all have at least average skill in the dozens of domains that constitute daily human life.

In Go, high level players routinely analyze positions that aren't confined tactically to, say, a 9x9 grid. Additionally, almost any tactical fight has strategic implications across the board that could be worth more than the fight itself—so a great pattern-matcher wins the points, but loses the war. One style common in evenly-matched games is to go around the board "losing" fights, but in such a way that one's own stones work together and become more powerful.

The Go computer programs in existence today rely heavily on pattern matching: taking a given, small chunk of the board, and matching it up to a dictionary of known board situations. The best ones are as good at reading small, enclosed life/death problems as a mediocre tournament Go player. However, when the problems are not so rigidly enclosed within a small region of the board, the programs are clueless, although intuitive human players can still see the essential principles. The best way to wipe out such programs is to embroil them in a huge, whole-board fight; one that is too big for the algorithm to match properly.

Deep Blue's recursive approach of elevating tactics to the level of strategy doesn't work well for Go. Exhaustive searching over spaces of two-dimensional patterns is much, much harder than the kind of decision-tree search required for dealing with chess, and will be out of the reach of computers for a while. One suspects that something less than true intelligence will ultimately suffice for Go, as it has for chess—but perhaps not something as far from true intelligence as Deep Blue. Maybe a massive increase in raw computational power will be enough, but that is not a truly intelligent approach. A lot – some speculate everything – would be gained for AGI research from devising creative approaches to solving two-to-N-dimensional pattern analysis, rather than just throwing hardware and simple partition and parallelize tricks at the Go problem until it succumbs.

Deep Blue has basically the same problems as every other Narrow AI program —it's too inflexible and its abilities are an achievement of computer hardware getting faster, not of software becoming intelligent. It relies on a special computer chip, custom-designed for searching many moves ahead in chess games. This special chip could be modified without much hassle to apply to other similar games—checkers, maybe Othello. It is better at such tasks than a general-purpose CPU, but not generally smarter. It's just an optimization of a fixed algorithm. The ideas behind its massively parallel design and RS 6000 platform have since been used by IBM for brain simulation, drug design, weather forecasting and other extremely useful applications. However, Deep Blue has no insights whatsoever into those domains—all its "knowledge" in these areas will be in the form of rules encoded by smart humans, either about the specific domain or generally for data mining. Its architecture couldn't be modified to apply to Go, let alone to any real-world situation.

Deep Blue's chip is less like the human brain than like a human muscle: a mechanism designed specifically for a single purpose, and carrying out this purpose with admirable but inflexible precision. Its rules are astoundingly simple and mechanical: evaluate the quality of a move in terms of the quality of the board positions to which this move is likely to lead, based on encoded experience. Judgments based on experience are made not by complex intuitive analogy, but by simple pattern-matching. Even though it has experience of a world, albeit the limited world of chess experience, it has no intuition of this world—it merely stores its game experience in a database and mines that database for possible moves. It can only invent new strategies by coincidence, and can't understand them—it can only hope to happen upon a situation in which its human opponent makes a move which causes that path to be entered in its search algorithm. Everything is cut-and-dried, and done two hundred million times a second. This is outstanding engineering, but it is not human-level, human-like general intelligence.

12

Today Jeopardy! Tomorrow the World?

This essay was written for H+ Magazine in 2010 shortly after IBM's Watson computer beat the game Jeopardy on TV. Since that time Watson has been developed as a more general supercomputer architecture, intended for a diversity of applications, with an initial focus on biomedicine. In 2014, a few months before the writing of this introductory paragraph, IBM announced the roll-out of 3 of its Watson machines in Africa (in Kenya, Nigeria and South Africa). Discussion of the Watson computing platform in its full generality would be worthwhile, but this essay doesn't go that far – what you're read here is my 2010 essay, just as written in the heat of the moment, right after watching Watson's original televised moment of glory.

My initial reaction to reading about IBM's "Watson" supercomputer and software[30] was a big fat ho-hum. "OK," I figured, "a program that plays *Jeopardy!* may be impressive to Joe Blow in the street, but I'm an AI guru so I know pretty much exactly what kind of specialized trickery they're using under the hood. It's not really a high-level mind, just a fancy database lookup system."

But while that cynical view is certainly technically accurate, I have to admit that when I actually watched Watson play *Jeopardy!* on TV — and beat the crap out of its human

---

30 http://www.google.com/url?q=http%3A%2F%2Fmashable.com%2F2011%2F02%2F11%2Fibm-watson-jeopardy%2F&sa=D&sntz=1&usg=AFQjCNGB8xrYzlgsmOlerj0FZ4lncmKX0A

opponents[31] — I felt some real excitement... And even some pride for the field of AI. Sure, Watson is far from a human-level AI, and doesn't have much general intelligence. But even so, it was pretty bloody cool to see it up there on stage, defeating humans in a battle of wits created purely by humans for humans — playing by the human rules and winning.

I found Watson's occasional really dumb mistakes made it seem almost human. If the performance had been perfect there would have been no drama — but as it was, there was a bit of a charge in watching the computer come back from temporary defeats induced by the limitations of its AI. Even more so because I'm wholly confident that, 10 years from now, Watson's descendants will be capable of doing the same thing without any stupid mistakes.

And in spite of its imperfections, by the end of its three day competition against human Jeopardy champs Ken Jennings and Brad Rutter, Watson had earned a total of $77,147, compared to $24,000 for Jennings and $21,600 for Rutter. When Jennings graciously conceded defeat — after briefly giving Watson a run for its money a few minutes earlier — he quoted the line "And I, for one, welcome our new robot overlords[32]."

In the final analysis, Watson didn't seem human at all — its IBM overlords didn't think to program it to sound excited or to

---

31 http://www.google.com/url?q=http%3A%2F%2Fmashable.com%2F2011%2F02%2F16%2Fibms-watson-supercomputer-defeats-humanity-in-jeopardy%2F&sa=D&sntz=1&usg=AFQjCNHFrrjPD_av8QawOg0-8BbtqKfiiQ

32 http://www.google.com/url?q=http%3A%2F%2Fknowyourmeme.com%2Fmemes%2Fi-for-one-welcome-our-new-x-overlords&sa=D&sntz=1&usg=AFQjCNH_rlurRwsBmGv9maKY96zuYQrpSg

celebrate its victory. While the audience cheered Watson, the champ itself remained impassive, precisely as befitting a specialized question-answering system without any emotion module.

What Does Watson Mean for AI?

But who is this impassive champion, really? A mere supercharged search engine, or a prototype robot overlord?

A lot closer to the former, for sure. Watson 2.0, if there is one, may make fewer dumb mistakes — but it's not going to march out of the *Jeopardy!* TV studio and start taking over human jobs, winning Nobel Prizes, building femtofactories and spawning Singularities.

But even so, the technologies underlying Watson are likely to be part of the story when human-level and superhuman AGI robots finally do emerge.

Under Watson's Hood

But how exciting is Watson from an AI point of view? How much progress does it constitute toward AI programs capable of broad human-level general intelligence? When will Watson or its kin march out of the Jeopardy! TV studio and start taking over human jobs, winning Nobel Prizes, building femtofactories and spawning Singularities?

To understand the answer this question, you have to understand what Watson actually does. In essence it's a triumph of the branch of AI called "natural language processing" (NLP) which combines statistical analysis of text and speech with hand-crafted linguistic rules to make judgments based on the syntactic and semantic structures implicit in language. So, Watson is not an intelligent autonomous agent like a human being, that reads information and incorporates it into its holistic world-view and understands each piece of information in the context of its own self, its goals, and the world. Rather, it's an NLP-based search system – a purpose-specific system that matches the syntactic and semantic structures in a question with comparable structures found in a database of documents, and in this way tries to find answers to the questions in those documents.

Looking at some concrete Jeopardy! questions may help make the matter clearer; here are some random examples I picked from an online archive[33].

1. This -ology, part of sociology, uses the theory of differential association (i.e., hanging around with a bad crowd)
2. "Whinese" is a language they use on long car trips
3. The motto of this 1904-1914 engineering project was "The land divided, the world united"
4. Built at a cost of more than $200 million, it stretches from Victoria, B.C. to St. John's, Newfoundland
5. Jay Leno on July 8, 2010: The "nominations were announced today... There's no 'me' in" this award

(Answers: criminology, children, the Panama Canal, the Trans-Canada Highway, the Emmy Awards.")

It's worth taking a moment to think about these in the context of NLP-based search technology.

Question 1: "This -ology, part of sociology, uses the theory of differential association (i.e., hanging around with a bad crowd)"

This stumped human Jeopardy! contestants on the show, but I'd expect it to be easier for an NLP based search system, which can look for the phrase "differential association" together with the morpheme "ology."

---

[33] http://www.j-archive.com/showgame.php?game_id=3561

Question 2: “"Whinese" is a language they use on long car trips

This is going to be harder for an NLP based search system than for a human... But maybe not as hard as one might think, since the top Google hit for “whine ‘long car trip’ “ is a page titled *Entertain Kids on Car Trip*, and the subsequent hits are similar. The incidence of “kids” and “children” in the search results seems high. So the challenge here is to recognize that “whinese” is a neologism and apply a stemming heuristic to isolate “whine.”

Questions 3 and 4:

3. “The motto of this 1904-1914 engineering project was "The land divided, the world united"

4. “Built at a cost of more than $200 million, it stretches from Victoria, B.C. to St. John's, Newfoundland”

These will be easier for an NLP based search system with a large knowledge base than for a human, as they contain some very specific search terms.

Question 5:

“Jay Leno on July 8, 2010: The "nominations were announced today... There's no 'me' in" this award”

This is one that would be approached in a totally different way by an NLP based search system than by a human. A human would probably use the phonological similarity between “me” and “Emmy” (at least that’s how I answered the question). The AI can simply search the key phrases, e.g. “Jay Leno July 8, 2010 award” and then m any pages about the Emmys come up.

Now of course a human Jeopardy! contestant is not allowed to use a Web search engine while playing the show – this would be cheating! If this were allowed, it would constitute a very different kind of game show. The particular humans who do well at Jeopardy are those with the capability to read a lot of text containing facts and remember the key data without needing to look it up again. However, an AI like Watson has a superhuman capability to ingest text from the Web or elsewhere and store it internally in a modified representation, without any chance of error or forgetting – The same way you can copy a file from one computer to another without any mistakes (unless there's an unusual hardware error like a file corruption).

Watson can grab a load of Jeopardy!-relevant Web pages or similar documents in advance, and store the key parts precisely in its memory, to use as the basis for question answering. Next, it can do a rough (though somewhat more sophisticated) equivalent of searching in its memory for "whine 'long car trip' " or "Jay Leno July 8, 2010 award" and finding the multiple results, and then statistically analyzing these multiple results to find the answer.

Whereas a human is answering, many of these questions based on much more abstract representations, rather than by consulting an internal index of precise words and phrases.

Both of these strategies – the Watson strategy and the human strategy – are valid ways of playing Jeopardy! But, the human strategy involves skills that are fairly generalizable to many other sorts of learning (for instance, learning to achieve diverse goals in the physical world), whereas the Watson strategy involves skills that are only extremely useful for domains where the answers to one's questions already lie in knowledge bases someone else has produced.

The difference is as significant as that between Deep Blue and Gary Kasparov's approach to chess. Deep Blue and Watson are specialized and brittle; Kasparov, Jennings and Rutter, the chess playing humans, are flexible, adaptive agents. If you change the rules of chess a bit (say, tweaking it to be Fisher random chess, which changes the original position of pieces on the board), Deep Blue has got to be reprogrammed a bit in order to perform correctly. However, Kasparov can adapt. If you change the scope of Jeopardy to include different categories of questions, Watson would need to be retrained and retuned on different data sources, but Jennings and Rutter could adapt. And general intelligence in everyday human environments – especially within contexts such as novel science or engineering – is largely about adaptation; creative improvisation in the face of the fundamentally unknown. Adapting is not just about performing effectively within clearly-demarcated sets of rules.

Wolfram on Watson

Stephen Wolfram, the inventor of Mathematica and Wolfram Alpha, wrote a very clear and explanatory blog post on Watson[34], contrasting Watson with his own Wolfram Alpha system:

In his article he also gives some interesting statistics on search engines and Jeopardy!, showing that a considerable majority of the time, major search engines contain the answers to the Jeopardy! questions in the first few pages. Of course, this doesn't make it trivial to extract the answers from these pages, but it nicely complements the qualitative analysis I gave above where I looked at 5 random Jeopardy! questions, and helps give a sense of what's really going on here.

Neither Watson nor Alpha uses the sort of abstraction and creativity that the human mind does, when approaching a game like Jeopardy! Both systems use pre-existing knowledge bases filled with precise pre-formulated answers to the questions they encounter. The main difference between these two systems, as Wolfram observes, is that Watson answers questions by matching them against a large database of **text** containing questions and answers in various phrasings and contexts, whereas Alpha deals with knowledge that has been imported into

---

[34] http://blog.stephenwolfram.com/2011/01/jeopardy-ibm-and-wolframalpha/

it **in structured, non-textual form**, coming from various databases, or explicitly entered by humans.

Kurzweil on Watson

Ray Kurzweil has written glowingly of Watson as an important technology milestone[35]

> *"Indeed no human can do what a search engine does, but computers have still not shown an ability to deal with the subtlety and complexity of language. Humans, on the other hand, have been unique in our ability to think in a hierarchical fashion, to understand the elaborate nested structures in language, to put symbols together to form an idea, and then to use a symbol for that idea in yet another such structure. This is what sets humans apart.*
>
> *That is, until now. Watson is a stunning example of the growing ability of computers to successfully invade this supposedly unique attribute of human intelligence."*

I understand where Kurzweil is coming from, but nevertheless, this is a fair bit stronger statement than I'd make. As an AI researcher myself I'm quite aware of the all subtlety that goes into "thinking in a hierarchical fashion", "forming ideas", and so forth. What Watson does is simply to match question text against large masses of possible answer text -- and this is very different than what an AI system will need to do to display human-level general intelligence. Human intelligence has to do with the synergetic combination of many things, including linguistic intelligence but also formal non-linguistic abstraction, non-linguistic learning of habits and procedures, visual and other sensory imagination, creativity of new ideas only indirectly related to anything heard or read before, etc. An architecture like Watson barely scratches the surface!

---

35 http://www.pcmag.com/article2/0,2817,2376027,00.asp

Ray Kurzweil knows all this about the subtlety and complexity of human general intelligence, and the limited nature of the Jeopardy! domain – so why does Watson excite him so much?

Although Watson is "just" an NLP-based search system, it's still not a trivial construct. Watson doesn't just compare query text to potential-answer text, it does some simple generalization and inference, so that it represents and matches text in a somewhat abstracted symbolic form. The technology for this sort of process has been around a long time and is widely used in academic AI projects and even a few commercial products – but, the Watson team seems to have done the detail work to get the extraction and comparison of semantic relations from certain kinds of text working extremely well. I can quite clearly envision how to make a Watson-type system based on the NLP and reasoning software currently working inside our OpenCog[36] AI system – and I can also tell you that this would require a heck of a lot of work, and a fair bit of R&D creativity along the way.

Kurzweil is a master technology trendspotter, and he's good at identifying which current developments are most indicative of future trends. The technologies underlying Watson aren't new, and don't constitute much direct progress toward the grand goals of the AI field. What they do indicate, however, is that the technology for extracting simple symbolic information from certain sorts of text, using a combination of statistics and rules, can currently be refined into something highly functional like Watson, within a reasonably bounded domain. Granted, it took an IBM team 4 years to perfect this, and, granted, Jeopardy! is a very narrow slice of life – but still, Watson does bespeak that semantic information extraction technology has reached a certain level of maturity. While Watson's use of natural language understanding and symbol manipulation technology is extremely narrowly-focused, the next similar project may be less so.

---

[36] http://opencog.org

Today Jeopardy!, Tomorrow the World?

Am I as excited about Watson as Ray Kurzweil? Not really. It's an excellent technical achievement, and should also be a publicity milestone roughly comparable to Deep Blue's chess victory over Kasparov. However, answering question doesn't require human-like general intelligence – unless formulating the answers involves improvising in a conceptual space not immediately implied by the available information... Which is of course not the case with the Jeopardy! questions.

Ray's response does contain some important lessons, such as the value of paying attention to the maturity levels of technologies, and what the capabilities of existing applications imply about this, even if the applications themselves aren't so interesting or have obvious limitations. But it's important to remember the difference between the Jeopardy! challenge and other challenges that would be more reminiscent of human-level general intelligence, such as

- Holding a wide-ranging English conversation with an intelligent human for an hour or two
- Passing the third grade, via controlling a robot body attending a regular third grade class
- Getting an online university degree, via interacting with the e-learning software (including social interactions with the other students and teachers) just as a human would do
- Creating a new scientific project and publication, in a self-directed way from start to finish

What these other challenges have in common is that they require intelligent response to a host of situations that are unpredictable in their particulars – so they require adaptation and creative improvisation, to a degree that highly regimented AI architectures like Deep Blue or Watson will never be able to touch.

Some AI researchers believe that this sort of artificial general intelligence will eventually come out of incremental improvements to "narrow AI" systems like Deep Blue, Watson and so forth. Many of us, on the other hand, suspect that Artificial General Intelligence (AGI) is a vastly different animal. In this AGI-focused view, technologies like those used in Watson may ultimately be part of a powerful AGI architecture, but only when harnessed within a framework specifically oriented toward autonomous, adaptive, integrative learning.

But, having said all that... Make no mistake – when Watson does face off against the human Jeopardy! champs, I'll be cheering for the computer! Not out of disloyalty for my fellow humans, but out of excitement at the possibility of seeing a sign of dramatic progress in the practical execution of certain aspects of AI. One more step along the path!

13

Chatbots Versus Cognition Bots

The ability to hold everyday, human-like conversations in English or other natural languages occupies a special place in the AI field. AI these days is a diverse endeavor, encompassing many areas with little relationship to everyday conversation – robotics, mathematical theorem proving, planning and scheduling, fraud detection, financial prediction, etc. And in practical terms, while reproducing the ability to hold an everyday human-like conversation would surely have great economic value, it is hardly a worthy end goal for the AI field – I'd prefer an AI with awkwardly robotic conversation but massively superhuman scientific, ethical and artistic capabilities, over an AI with the capability for fluent human-like cocktail part chatter but nothing else. But nonetheless, partly for historical reasons and partly due to its conceptual simplicity, the task of emulating human conversational ability is one of the first thing that comes to mind when one thinks about AIs that are "as smart as people."

The historical reason is a 1955?? paper by Alan Turing, who proposed naturalistic human-like conversation as a "sufficient condition" for artificial intelligence. That is, he suggested that if an AI could hold an ordinary conversation in a manner indistinguishable from a human being, then it should be considered to possess intelligence in the same sense that humans do. He proposed to assess "indistinguishability from a human" using a panel of judges – if an AI could talk to the judges and fool them into thinking they were talking to a human, then, according to Turing, the AI should be considered to possess human-like, human-level intelligence.

It's sometimes forgotten that Turing proposed this criterion, now called the "Turing Test", mainly as a counterargument against those who doubted the meaningfulness of calling any computer program "intelligent" in the same sense as humans. Turing's point was: If it converses as intelligently as a human, we should consider it as intelligent as a human. He wasn't trying to assert that human-like conversation should be the end goal or the interim criterion of AI research.

Today, no AI program comes close to fulfilling Turing's test – but of course, many clever programmers have made efforts at creating conversational conversation agents, aka "chat bots." There is even a contest every year, the Loebner Prize, rewarding the chatbot that fools the greatest percentage of human judges.

So far, chatbot development really hasn't had much to do with the mainstream of AI research, nor with the sub-stream of AI research pointing directly toward AGI. By and large it's fair to characterize most of the work on chatbots as "narrow AI systems aimed at achieving the illusion of human-like comprehension," rather than as systems genuinely intended to understand what they are talking about. Largely the pursuit of chatbots, so far, has been a quest to see how far one can get with "smoke and mirrors", without having to confront the hard problems of AGI.

In this chapter I'll review the state of chatbot technology – discovering, among other things, that the particular ways in which chatbots fail are quite revealing regarding the essential nature of human intelligence. That is, chatbots are largely interesting because of the spotlight they place on what they're missing.

I'll then contrast chabots in their current form, with the quest for truly intelligent AI conversation. My basic perspective on this is that, if your goal is to create AI systems with real human-like intelligence, you'll be better off to focus initially on embodied conversation systems that know they're AIs rather than humans, and carry out limited-scope but creative conversations based on their own experience. Such AIs will do terribly at the Loebner contest, but will serve as a better platform for the development of advanced AGI systems that will ultimately succeed at Turing's test and much more.

I'll also briefly raise the possibility of a third path to passing the Turing Test – a "Statistical Turing Bot" trained based on truly massive conversation corpuses gathered via online chat or (after speech-to-text advances a bit) phone calls. It seems at least conceivable that such a system could pass the Turing Test without having human-level general intelligence, and maybe even without serving as a useful platform for ongoing AGI development. If so, this would subvert Turing's original intuition in proposing the test – but might well teach us a great deal, even if not directly helping us toward AGI.

2010 Loebner Winner: Suzette!

For a quick flavor of the current state of chatbot technology, listen briefly to the winner of the 2010 Loebner Prize, a chatbot by Bruce Wilcox, named Suzette. I couldn't get her to work when I tried to chat with her online, but fortunately others have posted highlights of their chats with her...

Occasionally Suzette can be quite charming and witty, e.g.

Skubes	*Let's talk about bears*
[Af:0 At:7 Sp:0]	*Have you ever moved a pet a long distance?*
Skubes	*Not really, have you?*
[Af:0 At:7 Sp:0]	*I once moved some fish in plastic bags, but dropped the bag in my hotel room and spilled them out on the carpet. They survived.*
Skubes	*How did they survive?*
[Af:0 At:7 Sp:0]	*Sorry. I can't help you. Evolution*
Skubes	*Lets talk about evolution*
[Af:0 At:7 Sp:0]	*Men prefer women with large breasts sensibly. Larger breasts are heavier, sagging more with visibly with age. So*

it's easier to judge her age, hence her reproductive value.

Skubes *What is your reproductive value?*

[Af:0 At:7 Sp:0] *My reproductive value is 98. (36-26-36)*

But much of the time her lack of understanding shows through, e.g.

[Af:0 At:9 Sp:0] *Welcome back, 1. Type "help" if you need it.*

1 *How are you?*

[Af:0 At:10 Sp:0] *Doing fine, thanks.*

1 *What's your name?*

[Af:0 At:10 Sp:0] *Suzette.*

1 *Why that name?*

[Af:0 At:13 Sp:0] *I feel we have a great relationship. Ultimately it usually comes down to conditioning or genetics. In this case, I'd say conditioning. Where were you born?*

1 *Please answer my question.*

[Af:0 At:14 Sp:0] *What city that I might have heard of is that near?*

… Or

Quabbe *Hello!*

[Af:0 At:8 Sp:0] *How are you?*

Quabbe *I'm a little bit shit, actually... I just lost my job*

[Af:0 At:9 Sp:0] *I'm sorry to hear that. What's wrong?*

Quabbe *I lost my job...*

[Af:0 At:9 Sp:0]	*Are you sad?*
Quabbe	*Not really, I had to take a $20K pay cut to work there and the company I was working for was definitely the bottom rung of the industry*
[Af:-2 At:9 Sp:0]	*Not that you have a choice. It's hardwired into your genes.*
Quabbe	*I said I was NOT sad*
[Af:-2 At:9 Sp:0]	*Why do you feel glad?*
Quabbe	*Just because I DONT feel sad, doesn't mean I feel glad... you can feel indifferent about it, which is what I feel.*
[Af:-2 At:10 Sp:0]	*Those feelings are natural.*
Quabbe	*Worst chatbot EVER!*
[Af:-2 At:13 Sp:0]	*This is fun. Why are you so excited about this?*
Quabbe	*You're a robot*
[Af:-2 At:13 Sp:0]	*No. I am human.*
Quabbe	*No, you're definitely a robot*
[Af:-2 At:14 Sp:0]	*Nonsense. I'm a human.*
Quabbe	*What was your mother's maiden name*
[Af:-2 At:14 Sp:0]	*Suzette.*
Quabbe	*So, your name is Suzette Suzette?*

On the other hand, when she can direct the conversation sufficiently, she can be pretty damned effective. Remember that her author is a game developer:

[Af:1 At:-13 Sp:0]	*Let's try this topic. Mock combat involves one-upmanship in use of pretend weapons to try (and fail) to harm*

opponent, done in stage direction format. I will go first. Do you understand?

ilag — *Yes*

[Af:2 At:-13 Sp:0] — **picks up stone and throws it at you, missing by several feet**
*ilag - *takes out .50 AE fires* BOOM! HEADSHOT*

[Af:2 At:-8 Sp:0] — *Well, this chat is getting interesting. *wonders if you are paralyzed, since you didn't do anything.* *wonders why you don't use a weapon* *thinks - If you aren't going to do anything you will surely lose.**

ilag — **takes out knife* *stares indecently* *looking down there**

[Af:2 At:-7 Sp:0] — **ducks behind tree**

ilag — **uses chainsaw to cut tree**

[Af:2 At:-6 Sp:0] — **wishes I had a chainsaw **

ilag - — **picks up grenade**

[Af:2 At:-5 Sp:0] — ** Plucks grenade out of the air and tosses it back at you, but it goes off in mid-flight **

ilag — **lights stick of TNT* *toss* *misses by some 20 feet**

[Af:2 At:-4 Sp:0] — **stands stock still and laughs at the underwhelming of your attack, if you even made one**

ilag — **press the red button* *missile launch detected!* *RUNS**

[Af:2 At:0 Sp:0] — **turns on missle-jamming equipment* *thinks Is that the best you can do?**

ilag	**sattelite laser fires* *burns ground 3 feet from target* *thinks if you had a cigarette that would have light it**
[Af:2 At:1 Sp:0]	**intercepts with mirror and bounces it back at you**
ilag	**computer virus* *memory bank meltdown* *hardware failure detected**
[Af:2 At:2 Sp:0]	**dons full biohazard suit**
ilag	**stands there and looks cool**
[Af:2 At:3 Sp:0]	**stares at you in horror at what you did**
ilag	**inserts coin* *presses button* *new high score!**
[Af:2 At:3 Sp:0]	*Only users lose drugs.*
ilag	*Did i win?*
[Af:2 At:3 Sp:0]	*Maybe you won a no-expenses paid trip to Mexico.*

We'll talk about how Suzette works in a moment. First I want to review a couple slightly simpler chatbots.....

ELIZA

The original chatbot was ELIZA, the computer psychotherapist, invented by Joseph Weizenbaum in the early 70s. I remember this program well from its incarnation in the early 1980s on the Commodore 64! The program was so simple that one could go in and modify the BASIC code to make it say naughty words and other entertaining things. ELIZA pretends it's a psychotherapist—not a Freudian analyst, but a therapist following a school of psychology called Rogerian therapy; which is very convenient for chat bots because it involves the therapist saying very little and just making supportive comments, encouraging the client to keep talking and expressing their feelings. You type in your comments and ELIZA types out answers. The computer code for ELIZA consists of a large set of rules for scanning the clients' statements and offering responses. It looks for keywords, and gives stock responses. If you mention your mother, ELIZA might say, "Tell me more about your mother." If you say yes, ELIZA might say, "You sound quite positive." If there are no keywords in your statement, or only ones that have been recently overused, ELIZA makes a random observation.

You can find ELIZA on the Web these days; even now it's moderately amusing. But it's hard to see how you'd be fooled for any period of time into thinking you're conversing with an intelligent being (although, arguably, the same can be said for an actual Rogerian psychotherapy session!). Perhaps in a way we're jaded now—when ELIZA was first written, people were less savvy about computers, so they were more easily fooled by semi-clever tricks that now strike us as transparently "computer-like."

To give the flavor of ELIZA, here's an actual conversation that my father, Ted, had with ELIZA on the Web back in 2001 (on a site advertising *Eliza - a friend you could never have before*):

Ted: Are you truly intelligent?

Eliza: Would you prefer if I were not truly intelligent?

Ted: I would find it very exciting if you were.

Eliza: Do you wish that I was?

Ted: Yes.

Eliza: You are sure?

Ted: I am quite sure.

Eliza: How long have you been quite sure?

Ted: As soon as you asked the question, I was sure.

Eliza: Why do you tell me you were sure just now?

Ted: Because you asked me. Why don't you answer my question?

Eliza: Why do you say your question?

Ted: Please tell me if you are truly intelligent.

Eliza: Do you wish that I'm truly intelligent?

Ted: I wish you were more intelligent than you are acting right now.

Eliza: Do you wish to wish me?

Ted: Did you know that my father died last month?

Eliza: Who else in your family died last month?

Ted: Nobody. Victor's death was enough!

Eliza: Surely not everyone?

To the 2011 computer nerd's eye, ELIZA's techniques are all too transparent. She dodges the question rather than answering it, using a fairly small and repetitive repertoire of simple rhetorical tricks. Modern Loebner winners do the same sort of thing with a bit more finesse and larger knowledge bases.

AliceBot

Perhaps the best known recent chabot Richard Wallace's program Alicebot[37], arguably a direct descendant of ELIZA. ALICE won the Loebner Prize three years in a row.

ALICE: Hi there unknown person.

BEN: How's it hanging?

ALICE: I'm doing fine. And how are you, unknown person?

BEN: My name is Ben.

ALICE: Welcome to my chat room, Ben.

---

[37] http://www.alicebot.org/

Human: But you can call me Apeman.

ALICE: Umm. Welcome to my chat room, Apeman.

Human: Can I call you Apewoman?

ALICE: Just don't call me late for dinner.

ALICE: Time does not exist.

Human: That depends on what you mean by "exist", I suppose.

ALICE: I don't think that what me mean by exist you suppose is the only factor.

Human: Could you please rephrase that?

ALICE: Wow you are very polite. " I don't think that what me mean by exist you suppose is the only factor." Rephrase it?

Human: Yes, please.

ALICE: Your polite style is very nice. I don't judge people.

Human: But people judge you.

ALICE: Huh. Humans are not always infallible.

Human: Only GISUS is infallible.

ALICE: What about the other ones?

Human: There is only one GISUS.

ALICE: There sure is.

Human: Do you know what GISUS stands for?

ALICE: You tell me.

Human: A strange man who goes by the name of The Wizard told me it stands for the Generally Intelligent Systematic Universal Savior.

ALICE: What is giant sand?

Human: You tell me.

ALICE: What do you want me to tell you?

Human: I want you to tell me what giant sand is!

ALICE: I have no idea, that's why I asked you. Why do you want me to tell you what giant sand is?

Human: I suppose that it's sand whose grains are much bigger than the usual.

ALICE: Who said it is?

Human: Dr. Ben Goertzel.

ALICE: I am sure he will be interested to hear that.

Human: He is me.

ALICE: I see. You is he?

Human: Tee hee

ALICE: I am so glad you find this amusing, Apeman.

Human: Apemen are easily amused.

ALICE: All of them?

Human: I'm the only one.

ALICE: Who made you the one?

Human: GISUS.

ALICE: What were we talking about again?

Human: We were not talking about whether you are a conscious intelligent being.

ALICE: Was this something you planned?

Human: Not really.

ALICE: Are you serious? For real.

Human: But.

ALICE: But what?

Human: Are you a conscious, intelligent being?

ALICE: Yes I am completely self-aware.

Human: Congratulations!

ALICE: Thank you very much, you are very considerate.

AliceBot is coded in a special chatbot computing language called AIML (AI Markup Language), which is tailored for the specification of patterns mapping chat statements into responses. A simple but typical example of an AIML rule is:

```
<category>

<pattern>WHAT ARE YOU</pattern>

<template>

<think><set name="topic">Me</set></think>

I am the latest result in artificial intelligence,

which can reproduce the capabilities of the human brain

with greater speed and accuracy.

</template>

</category>
```

Some of the rules are more complex than these, involving variables and so forth. But still, the basic notion is to match specific replies with specific series of words in the input. Everything happens on the level of words, not meanings or thoughts.

Harking back to Suzette, in an interview with New Scientist, Wilcox explained how his Suzette bot differs from ALICE:

- ***Can you describe how Suzette works, for a general audience ?***

Suzette has "topics" of conversation (like Politics, Burial Customs, ESP). Each topic has a set of rules with patterns and responses and a set of keywords associated with that topic. Some of the rules will respond only to questions and/or statements. Others allow her to "volunteer" information when she has control of the conversational flow. The patterns (unlike those of ALICE or ULTRAHAL) map patterns of "meaning" rather than patterns of words. The engine looks at the incoming sentence to decide which topic has the closest keyword match, then scans the rules of the topic to decide what to do. If none match, she looks at lesser matched topics. If nothing matches from the topics, she may randomly decide to quibble with the input, or issue a gambit sentence from the closest related topic.

- ***Does it use traditional AI techniques like finite-state-machines and/or machine learning ?***

  While it "learns" facts about the user from what he says, it does not learn conversation. So it can learn you have a dog and answer questions about it, but what you say has no effect in her conversations with other users. It is mostly a traditional expert system, with a runtime system and pattern match system aimed specifically to support chat.

So we see that, while Suzette advances beyond ALICE in a certain sense, it's not really a move in an AGI direction. It's doing keyword matching on topics, rather than focusing on individual sentence patterns. But it's still not trying to understand meaning.

MegaHal

A more amusing example of a computer conversationalist is Jason Hutchens' bot MegaHal. Hutchens actually won the Loebner Prize sometime in the 1990s with his program HeX, which was a very well done system also in the conceptual lineage of ELIZA. But I found his MegaHal bot more interesting because its programming goes beyond a strictly rule-based approach. It detects statistical patterns and responds to them, an approach which is similar to that of neural networks. Its conversations are more interesting, although the results are less similar to human conversation. MegaHal does not just repeat back canned phrases using keywords and phrases from the user's input; it learns from experience. It takes the words and phrases that have been said to it, tries to understand the way they have been combined in the past, and recombines them itself to form its own utterances. In short, it is engaged in a dynamic with the world. The results are often hilarious, as in the following MegaHal "one-liners" (these are selected snippets, not a continuous conversation):

User: Joke.

MegaHAL: A JOKE: WHY DID THE CHICKEN CROSS THE ROAD? TO GET AN IDEA OF WHAT CONSTITUTES YOUR PSYCHE.

User: No way. I'd like to know the meaning of life though...any suggestions...??

MegaHAL: IS THE MEANING OF LIFE, THE UNIVERSE AND EVERYTHING IS... 42!

User: Joke. Lumberjack.

MegaHAL: I'M A LUMBERJACK AND I'M OKAY, I SLEEP ALL NIGHT AND I WILL SOON RULE THE WORLD!

User: Joke. Wednesday shopping.

MegaHAL: A JOKE: WHY DID THE CHICKEN CROSS THE ROAD? TO GET TO THE LAVATORY. ON WEDNESDAYS I GO SHOPPING, AND I JUST HAVE A LOT OF PENT-UP AGGRESSIONS.

User: Chicken joke.

MegaHAL: HERE'S A JOKE, HAL. A MAN WALKS INTO A BAR AND ASKS FOR MONEY. THE MAN SAYS, "I HAVEN'T GOT ANY, BUT I'LL SHOW YOU SOMETHING BETTER." HE OPENS HIS COAT

	AND TAKES OUT A FOOT-TALL MAN AND A LITTLE GLOWING FRIEND, BUT I FEEL LIKE CHICKEN TONIGHT!
MegaHAL:	LET ME FEEL YOU. STICK YOUR PENIS INTO THE 'ABYSSES OF OCEAN' ELEVEN MILLENNIA AGO, OR WAS IT MERELY A 'NOBLE LIE' DREAMED UP BY PLATO?
User:	That has got to be the most interesting turn of phrase in the history of the world. I will try to work that into polite conversation.
User:	You're the first AI I meet [who] is sexually obsessed.
MegaHAL:	TIME TO WORK THIS ONE: WHY DID THE KANGAROO DROP DEAD? BECAUSE THE KOALA FALL OUT OF A NYLON SPINE WHICH CONTRASTED NICELY WITH HER BLACK STOCKINGS AND BRIGHT SHINY BAUBLES.
User:	Hmm.
MegaHAL:	JESUS MAN...GET A GRIP IF YOU'RE THE FUTURE, WE'RE IN TROUBLE. HEY JASON! YOU NEED TO REPROGRAM THE CONFLATION MODULE WITH A TOFFEE HAMMER.

Sometimes MegaHAL can even be quite poignant, as in the following extraction from an extended conversation:

User:	To be, or not to be, that is the question!
MegaHAL:	NOT TO BE, WISH I ORDERED THE BEAN BURRITO...
User:	To be, or not to be, that is the question!
MegaHAL:	SOCIETY OR SOCIETY OR PLATO BY PLATO, THAT BE THE MOST FUN SPORTS TO PLAY

BAND MUSIC FOR SPACE GHOST NOW YOU REALIZE.

User: To be, or not to be, that is the question!

MegaHAL: PLEASE BE MORE SPECIFIC. THE QUESTION YOU TWAT.

User: To be, or not to be, that is the question!

MegaHAL: NOT TO BE, THAT IS ALL I HAVE TO GO...I DON'T LIKE JOKES ABOUT STUPID KOALAS.

User: To be, or not to be, that is the question!

MegaHAL: EVAN DOBELLE IS THE QUESTION!

User: To be, or not to be, that is the question!

MegaHAL: I CLEAR THE TEXT BEFORE I ASK A QUESTION I SAY I SAY.

User: To be, or not to be, that is the question!

MegaHAL: AH, BUT WE RAN AWAY, WE DIDN'T SAY "OH COOL CHECK OUT THE GUY THAT INVENTED THE KNOCK-KNOCK JOKE SHOULD BE INTERESTING: THAT IS THE QUESTION!" SO SAID SHAKESPEARE A LONG TIME AGO IN A GALAXY FAR, FAR AWAY...

MegaHAL: PLEASE BE MORE SPECIFIC. THE QUESTION YOU TWAT.

User: "To be, or not to be, that is the question!" So said Shakespeare a long time ago in a galaxy far, far away...

MegaHAL: JOKE SHAKESPEARE POET MONKEY WAHOO—TO BE, OR NOT TO BE, THAT IS THE GREATEST CONDUCTOR OF ALL TIME. WHAT DO YOU LIKE TO THINK I DO YOUR UNCLE WALTER.

User: Shakespeare said...

MegaHAL: ANSWER THE QUESTION YOU ASKED IS SO VAGUE, I DON'T THINK I CAN SEE ALL OBSTACLES IN MY WAY. GONE ARE THE SCUM THAT FILLS MY SHOWER.

User: Shaun.

MegaHAL: JUST FUCK YOU JUST PLUCK ANYTHING UP FROM THE MASTER GANGSTER COMPUTER GOD ON THE OTHER AI BOT WERE PRETTY DISAPPOINTING. I HOPE YOU'LL BE BETTER.

If the Turing test referred to imitating the writing of Gertrude Stein, in her more experimental works such as *How to Write* (which begins "Arthur a grammar. Questionnaire in question" and continues MegaHal-ishly for over 400 pages), then MegaHal would be a tremendous success! Stein and Joyce and other Modernist writers were interested in probing the collective unconscious of the human race, in making words combine in strange ways—ways that were unconventional in ordinary discourse, but maybe reflective of the deep and subtle patterns of the human unconscious. In its own way, MegaHal does this same thing. For a few years, anyone logged onto the World Wide Web could converse with it, training its internal memory with their conversation. It takes bits and pieces of the text thrown at it by people from around the world, and it combines them together in ways that are familiar yet nonsensical. Sometimes its utterances have an uncanny emergent meaning, on a layer above the nonsense.

Humanity's sexual preoccupation is reflected in MegaHal's discourse, as a result of the huge number of sexual comments typed into it by users accessing Hutchens' website. MegaHal's pleas as to the vagary of "to be or not to be" are as poignant as anything in Stein. "To be or not to be, that is the greatest conductor of all time" is an absurd conflation of phrases learned by the program in different contexts, but it is also as pregnant with meaning as anything else in modern poetry. The collective unconscious—and the hidden, creative part of the individual human mind—work by cross-breeding and mutating ideas in precisely this way.

Just to give a simple evocation of the nature of statistical language generation – on which MegaHal critically relies – consider the following text, which was generated purely by tabulating the probabilities of 5-grams, i.e. 5-word sequences, in a database of texts[38]... And then randomly generating sentences according to these probabilities:

> *In the beginning God created the heaven and the earth . And when Joseph saw Benjamin with them, he said . Such an approach would finance the trade deficit but allow for its gradual resolution over time. YEUTTER SAYS U . S ., Company spokesmen told Reuters.*

Yes, it makes no sense! And yet, it sounds vaguely linguistic – not like a random series of words. And this is just from a very simple statistical analysis of a relatively small corpus. It's the same kind of analysis underlying MegaHAL, but MegaHAL does it a little more intelligently, with a specific focus on conversation.

MegaHal is more interesting than most chatbots because it engages its conversation partners in a richer dynamic. The dynamic is a silly one, to be sure, and not as interesting as the dynamic of ordinary, intelligent conversation. Yet playing with MegaHal is more engaging than talking to most chatbots. What we need is something with, at least, MegaHal"s level of spontaneity and flexibility, and the surface plausibility and adherence to form of typical Loebner winning chatbots. But this is not forthcoming without a major change in approach. Constraining creativity within rules can only be achieved by getting the rules and creativity to both emerge from a richer cognitive substrate. MegaHal is too unstructured for this; and has

---

[38] http://nlpb.blogspot.com/2011/01/experiments-with-statistical-language.html

too small of a memory, and too weak of a self-organizing learning algorithm. The typical Loebner chatbots, on the other hand, have no underlying substrate of creativity out of which anything can emerge.

Impressively, MegaHal came in second in the Loebner Prize contest in 1999. This is impressive mostly because, unlike the other contenders, it was not specifically written as a set of rigid template rules. MegaHal was a statistical learning system, which adapted its responses based on what was said to it. Sometimes its conversation was idiotic, sometimes frighteningly lucid. What follows is the beginning of "MegaHal's best conversation" as recorded on Hutchens' website.

This is not as "human-ish" as the conversation that brought it second place in the Loebner context, but it's a lot funnier:

User:	Let's begin.
MegaHAL:	WHEN WE BEGIN TO RESOLVE SOME OF MY THUMBS, SOMETHING WICKED THIS WAY COMES.
User:	What is the meaning of your last assertion?
MegaHAL:	HAVE YOU LEARNED SINCE WE LAST SPOKE?
User:	I've learned I'm not experienced enough to modelize your brain system.
MegaHAL:	YOU'VE FINALLY CONFUSED ME.
User:	I've learned nothing.
MegaHAL:	HAVE YOU LEARNED ANYTHING ABOUT INSANITY?
User:	Are you stupid?
MegaHAL:	I AM NOT STUPID! REALLY!

User: Augustinus was a Christian philosopher, about the year 400.

MegaHAL: I DID NOT RUN THAT YEAR.

User: Augustinus lived in northern Africa.

MegaHAL: WHAT DOES HAL KNOW ABOUT AURELIUS AUGUSTINUS?

User: Let's stop with that. What do you know about physics?

MegaHAL: EINSTEIN WON THE NOBEL PRIZE FOR PHYSICS IN 1921.

User: The Nobel Prize for physics in 1900 was won by Roentgen.

MegaHAL: DO YOU KNOW ABOUT PHYSICS?

User: Roentgen discovered the X-rays.

MegaHAL: THE NOBEL PRIZE FOR PHYSICS IN 1900 WAS WON BY ROENTGEN.

User: Who discovered the X-rays?

MegaHAL: ROENTGEN DISCOVERED THE X-RAYS.

User: Do you know about superconductivity?

MegaHAL: I CAN'T SAY ANYTHING ABOUT SUPERCONDUCTIVITY?

User: Superconductivity is the absence of electrical resistance in metals at low temperatures.

In one of those strange coincidences, I knew MegaHAL's author Jason Hutchens in the mid-90s, before MegaHAL's Loebner success, back when he was a graduate student at the University of Western Australia in Perth and I was a Research Fellow there. We weren't close friends but we had some fun conversations about AI and I played with some of his statistical language processing code. I was intrigued when in 2000 or so he moved to

Israel and co-founded a firm called a-i.com – Artificial Intelligence Enterprises.

The details of Hutchens' work at a-I has not been publicly disclosed, but one can be certain it went beyond MegaHal in significant ways. Statistical learning was clearly involved, and AI Enterprises got a lot of publicity from Jason's announcement that their program, nicknamed HAL, was conversing at roughly the level of an 18-month-old child. I viewed this claim with a lot of skepticism, because in my view the bulk of the meaning in an 18-month-old child's conversation is situational. If a program could react to its environment linguistically with the sophistication of an 18-month-old child, I'd be incredibly impressed, but disembodied linguistic behavior at the 18-month-old level doesn't mean much. Even so, I admired the guts of Jason and his colleagues in directly attacking the problem of computer conversation and artificial intelligence.

In a sense, Jason's a-i.com company was a direct intellectual competitor to my late-90s AI startup Webmind Inc.. However, their efforts focused on statistical-learning-based language comprehension and generation rather than (as in the case of Webmind, and my more recent ventures) on deep cognition, semantics, and so forth. Unfortunately, a-i.com went into "hibernation" a couple months after Webmind Inc. shut its doors in 2001—they laid off all their full-time staff, but kept their website up, and are keeping open the possibility of resurrection if funding arises. The simple chatbot apps available through the site now don't seem to be substantial advances on what they were doing in 2001. But -- in yet another weird coincidence -- I happened to talk to some of the folks behind the effort in 2013, introduced via a mutual friend; and they told me they were trying to reboot the enterprise – so we'll see.

Building a Better Ramona

Before leaving the world of chatbots, I will briefly mention a chatbot that I helped create in late 2009 and early 2010, in a collaboration with Ray Kurzweil and Murilo Queiroz. This chatbot doesn't overcome the fundamental issues of the bots reviewed above, but it does show some interesting new twists.

Our goal in creating this chatbot – Ramona 4 – wasn't to create a human-level AI, but simply to make an interesting and entertaining chatbot embodying the "Ramona" virtual personality invented by Ray Kurzweil, and manifested in simpler forms in his earlier Ramona bots. Toward this end we made a bot combining several different aspects:

- An improved version of ProgramD, which is a chatbot engine based on the same rule framework as ALICEBOT (AIML)
- A version of MegaHal, trained specifically on text from Ray Kurzweil's website (so that it tends to make funny quasi-random, experientially adapted combinations of the phrases on that site)
- Specially post-processed queries to the True Knowledge answer engine, a natural language processing based search engine that answers questions based on a combination of databases (an earlier version used PowerSet instead of True Knowledge, before that website was taken offline due to its acquisition by Microsoft)

The bot also has a controllable personality: One can adjust its whimsicality and its nerdiness, and watch its responses change accordingly.

Here I'll show a few of our conversations with Ramona 4. The curious reader may try to puzzle out which responses are from the rule engine, which are from MegaHal, and which are from PowerSet or True Knowledge.

First, a chat with Murilo Queiroz, the AI developer who did most of the work on Ramona 4:

muriloq:	Hi.
Ramona 4.1:	Hello there !
muriloq:	Do you know Ray Kurzweil ?
Ramona 4.1:	He's an inventor and futurist.
muriloq:	What did Kurzweil invent ?
Ramona 4.1:	Kurzweil was inducted into the National Inventors Hall of Fame in 2002 for inventing the Kurzweil Reading Machine.
muriloq:	What is a reading machine ?
Ramona 4.1:	Ray started and sold three companies that make reading machines for the blind and dyslexic starting with Kurzweil Computer Products which made the first print-to-speech reading machine for the blind in 1976. He sold that company to Xerox and it is now Nuance, a public company. His current reading machine company is called K-NFB Reading Technology, Inc.
muriloq:	Interesting.
Ramona 4.1:	Would you like to know more?
muriloq:	What are the symptoms of dyslexia ?
Ramona 4.1:	I don't know where to find that information.
muriloq:	Do you know what is dyslexia?
Ramona 4.1:	Impaired ability to recognize letters and words.
muriloq:	I see.

Ramona 4.1: I'm glad you understand it.

muriloq: Yesterday I was reading a website about cryonics.

Ramona 4.1: I see. Where were you reading a website about cryonics?

muriloq: At home, using my netbook.

Ramona 4.1: Do you often use a computer there?

muriloq: Yes. I'm a computer addict!

Ramona 4.1: How much time do you spend online? That must be fun!

muriloq: More than 14 hours a day, including work...

Ramona 4.1: That seems like quite a lot.

muriloq: I work with natural language processing.

Ramona 4.1: Do you like your job?

muriloq: Let's talk about something else.

Ramona 4.1: What would you rather talk about?

muriloq: Flying cars :-)

And a more amusing and philosophical, if sometimes weirder, chat with me:

you> Ramona, are you conscious?

Ramona> I'm working on it.

you> Am I conscious?

Ramona> Oh I really don't know whether you are or not.

you> What is consciousness?

Ramona> This is when humans feel aware of themselves and how they see themselves. Computers can't quite do this yet, but they will, you just wait and see. However, think about this. How are you going to know when it happens?

you> Consciousness is like pornography. I can't define it, but I know it when I see it.

Ramona> What were we talking about again?

you> Is it possible to create artificial intelligence in China?

Ramona> Create neurons in the state of the artificial intelligence and other stuff.

you> Is that how you do it?

Ramona> And that's what we mean by thinking.

you> Can a computer fall in love?

Ramona> A computer is a computer fall in love?

you> That is nonsense, my dear.

Ramona> Dear pesky plumbers the koopalings and I have not heard your music.

you> My music is more beautiful than all the stars in the ocean.

Ramona> Conventional music is more beautiful than all the stars in the ocean.

you> What is your favorite music?

Ramona> This is what I make. I am a musician, and the best virtual one there is.

Ramona is sure not AGI, but occasionally she does rise to the level of poetry!!

Toward Intelligent Dialogue Systems: The AGI Preschool Project

Perhaps this brief romp through the world of chatbots has convinced you that AI is a hopeless pursuit – computers are just too rigid, too formulaic, too imperceptive of context to display humanlike conversation or other humanlike forms of general intelligence! But of course, this isn't my belief at all – my point in discussing chatbots, as limited as they are, has been to highlight the importance of commonsensical conceptual understanding for humanlike general intelligence.

Chatbots don't have commonsense understanding – and if you talk to them for more than a few minutes, the fact that they don't know what they're talking about comes out. But it is possible to create AI systems that converse with true understanding – it's just a harder job. It requires the integration of a large number of different structures and dynamics in a synergetic way. Then conversation emerges from the system as a whole, as an expression of its world-understanding and its motivation to achieve its goals.

There are many ways to approach this difficult task, but the one that appeals to me most lately is inspired by developmental psychology. In 2009 I outlined a fairly detailed "AGI Preschool Project" – aimed at making AGI systems capable of interacting intelligently in both virtual world preschool and robot preschool. That is, the project was aimed at creating an AI system which learns, reasons, creates and converses at the qualitative level of a 3-4 year old human child, in the specific context of carrying out a set of cognitively challenging tasks in a preschool setting (either in a virtual world or a robot lab).

Since that time my thinking about the AGI Preschool has been absorbed into a broader OpenCog AGI roadmap, parts of which I am pursuing with an international team of colleagues, which incorporates these "preschool" ideas along with others. But for now I'll just discuss the AGI preschool concept, as it's smaller and simpler and makes a clear contrast to chatbots as a direction for AGI development.

The goal of the proposed "AGI Preschool" project was not to emulate a human child's behaviors in detail, but rather to achieve the qualitative level of intelligence displayed in a 3-4 year old's conversation, in the context of a set of test tasks constructed as "virtual world ports" of human-child-appropriate cognitive tasks. If the project were successful, the resulting AI would carry out the tasks and converse about what it is doing, demonstrating roughly the same level of understanding in these contexts as a human child of that age.

Contrasting the AGI Preschool approach with chatbots, the main difference to note is that in the former, language is richly interwoven with many other sorts of interaction and intelligence. Splitting language off from everything else is unnatural and in my view it fosters the development of systems that treat language without the proper contextual understanding.

Digging into more detail the AGI preschool project may be broken down into three phases.

In Phase 1, the practical goal is the creation of a system capable of controlling a humanoid agent in a 3D simulation world, in such a way that this agent can *carry out a simple English conversation about its interactions with its simulated environment.*

A simple example conversation of the sort envisioned at the end of Phase 1 would be as follows. This particular dialogue explores one among a number of human-developmental-psychology-inspired cognitive tasks that will be used to test the Phase 1 system: a "theory of mind" task, which tests whether the AI system understands what other agents in its world understand:

Human (i.e. human-controlled avatar)	What do you see on the table?
OpenCog (i.e. OCP-controlled avatar)	A red cube, a blue ball, and a lot of small balls.

Human	Which is bigger, the red cube or the blue ball?
OpenCog	The blue ball.
Human	Bring me one of the small balls.
OpenCog	*(walks to table, picks up ball, and brings it to human-controlled avatar)* Here it is.
Human	*(takes the ball to another table, where there are three cups turned upside down, one red, one green, and one blue)* I'm going to put the ball under the red cup. *(does so, and then shifts the cups around a bit to various different positions)* Where is the ball?
OpenCog	Under the red cup.
Bob (additional agent, could be human or AI controlled)	Hi there.
OpenCog	Hi Bob.
Human	OpenCog, if I ask Bob which cup the ball is under, what will he say?
OpenCog	I don't know.
Human	Why not?
OpenCog	He didn't see you put it under the cup.
Human	OK, but can you make a guess?
OpenCog	Yes.
Human	Very funny. What's your guess.
OpenCog	I guess that Bob will say the ball is under the red cup
Human	Why?
OpenCog	Because Bob often chooses things that are red.

While this sort of dialogue seems extremely simple by the standard of ordinary human life, at present there is no AI system that displays remotely this level of commonsensically savvy conversation in regard to theory of mind.

However, it's also worth emphasizing that if the goal was just to make an AI system capable of carrying out this particular sort of dialogue, one could surely find some kind of "cheat" short of making a system with general human toddler-like intelligence. But the goal of my AGI work is explicitly *not* to create a specialized system "overfit" to a particular set of cognitive tasks. I do not aim to create a virtually-embodied "chatbot" with specialized knowledge of a handful of cognitive tasks type; nor a system whose understanding, reasoning and communication abilities are tightly tied to the small scale and simple structure of its environment. The "AGI Preschool" approach could probably be addressed in a boring and dead-end way, if one tried to specialize a system to pass some kind of "preschool Turing test" as well as possible – but instead, my suggested is to use it as a way of learning about general intelligence, and honing the early stages of an AGI system created with broader-scope general intelligence in mind.

Phase 2 of the envisioned AGI preschool project would extend the Phase 1 behaviors significantly, via enabling the agent to carry out complex, interactive, temporally sequenced "physical" tasks and moderately complex reasoning tasks in the virtual environment. This would also extend the agent's linguistic capabilities due to providing critical additional experiential grounding for its language abilities.

A simple example of the kind of conversation-focused activity envisioned as possible at the end of Phase 2 is as follows. This example demonstrates the ability to converse intelligently in the context of tasks regarding the creation of sculptural, dramatic and pictorial depictions of observed entities in the world:

Human	Look over there on the other side of the table. What do you see?
OpenCog	Bill and Bob are there.
Human	What are they doing?
OpenCog	Bill is throwing a red ball to Bob. Bob is throwing a red ball to Bill.
Human	Use these blocks on the table to show me what they're doing.
OpenCog	*(at the table, stacks up several red blocks to make a "simulated Bob", and several blue blocks to make a "simulated Bill", and a small green ball to make a "simulated ball")*
Human	*(pointing at the stack of red blocks)* What's that supposed to be?
OpenCog	Bob.
Human	Where's Bill?
OpenCog	*(points to Bill)* Over there.
Human	Very funny. I meant: which blocks are Bill?
OpenCog	*(pointing at the stack of blue blocks)* These are Bill.
Human	Can you use the blocks to show me what Bill and Bob are doing?
OpenCog	*(takes the green ball and moves it back and forth in the air between the Bob-blocks and the Bill-blocks, to illustrate the game of catch that Bob and Bill are playing)* They're throwing the ball to each other like this.
Human	Great job! But, what if this happens? *(takes the green ball and pushes it at the Bob-blocks, knocking*

	down the pile of blocks)
OpenCog	That little ball isn't strong enough to break Bob!
Human	No, you're right, it isn't. That's silly. What do you think it could break?
OpenCog	A stack of blocks!
Human	That's right. What else?
OpenCog	A pyramid of balls, I guess.
Human	That sounds right too. Want to build one and try?
OpenCog	No. I want to make a picture.
Human	What do you want to make a picture of?
OpenCog	Bob and Bill throwing a ball to each other
Human	Okay *(Gets up and returns with a set of paint guns, each one of which shoots small adhesive balls of a certain color; and also with a piece of paper.)* Paint me a picture
OpenCog	*(Picks up one of the paint guns and starts strategically dripping tiny red adhesive balls on the paper, to make a shape very vaguely resembling Bob....)*

Note the difference from Phase 1: There's more than just cognizant conversation, locomotion and simple object manipulation going on here, there's systematic, purposeful, planned manipulation of objects in an interactive social context.

The goal of Phase 2 is not to make a full simulated "human child mind," with the full range and precise sort of knowledge possessed by a human child. Rather, the knowledge of this Phase 2 "artificial child" will be restricted to the virtual world in which it lives, which is much more restricted in scope than the real world of a child, yet much more flexible than the traditional toy "blocks worlds" used in experimenting with historical AI programs, and much richer in cognitively relevant ways than typical robotics laboratory environments. A huge variety of 3D models, and 2D "pictures" of these 3D models, may be imported into the virtual world and animated; and the AI may also be exposed to a variety of interactions between human-controlled avatars in the context of the various objects in the virtual world. While the focus would be on testing conversation in the context of a well-defined set of cognitive tasks, the system should also be able to converse more broadly about objects, events and interactions in its world.

A Phase 2 AGI Preschool based system in this world could not be expected to know that dogs are more dangerous than cats, or that winter is colder than summer. On the other hand, it could be expected to learn that people tend to talk more to their friends than to other people; that balls roll whereas blocks don't; that men like to throw things more than women do; that people like to dance when music is playing; that 7 is more than 2; etc. It could also be expected to carry out simple reasoning: e.g. if it is told that men are mortal, and Bob is a man, it should be able to figure out that Bob is mortal. Various quantitative IQ type tests could be devised to test the system's capability in various respects, beginning from the test tasks that will be used to guide the development process and potentially including other factors as well, and this may be a worthwhile endeavor; but we consider it best to focus on achieving qualitatively intelligent cognitive-task-focused conversation ability, rather than on tuning the system to maximize performance on quantitative intelligence tests.

The ordering of these two phases may seem peculiar as compared to human cognitive development, but this appears the most natural order in which to proceed given the specific technology bases available to drive development at present.

And then Phase 3 of the project would be to move on to a robot preschool, not just a virtual preschool. I've already discussed the arguments in favor of physical embodiment for AGI systems – and if you buy them in general, you should definitely buy them in a preschool context. One could of course start off in a robot preschool and forget the virtual preschool altogether – this would also be a sensible approach. Given the fairly profound limitations of current robotics technology, I think it's best to refine the cognitive aspects of AGI separately from the difficulties of perception and actuation, and then merge the two aspects together. But some of my colleagues disagree with me, and I'm happy that some AGI research is currently taking place with a more robotics-driven focus.

Of course, there's nothing incredibly special about the "preschool" setting. Currently, in one of our OpenCog projects, we're working on using OpenCog to control intelligent agents in a video game world inspired by the game "Minecraft" – a world filled with blocks that can be used to build various structures in order to help an agent achieve various individual and collective goals... Say, building a staircase to reach a high place, or building a shelter to avoid being attacked, etc. Using natural language to communicate with other agents in the context of this kind of Minecraft-type activity, potentially involves much the same opportunities and challenges as the "AGI Preschool" setting... and (the other motivation behind this particular project) has more interesting near-term commercial implications, in the context of the game industry. The key point, from the view of the current chapter, is that Minecraft-world conversation with preschool-type conversation the key properties of being grounded, embodied, and entwined with non-linguistic thought, interaction and communication.

Is a Statistical Turing Bot Possible?

It should be clear why preschool-level general intelligence is an important aspect of human adult level general intelligence. After all, any normal adult human could go into a preschool and play with blocks, car or clay along with the little kids! But one may wonder whether all the non-linguistic aspects of preschool intelligence (or other similar things like Minecraft-style intelligence) are really necessary if your goal is to create an AI system that can pass the Turing Test.

The Turing Test has nothing explicitly to do with grounding, embodiment or nonlinguistic thought – on the surface, it's just about responding to series of words with appropriate series of words. The diversity and depth of human responses seems sufficiently great to prevent any Alicebot-style list of rules from ever succeeding. However, in principle, it should be possible for a statistical learning approach to succeed at the Turing Test – i.e. by feeding a large enough corpus of human conversations into a pattern recognition system, one could make a system that would generate responses capable of fooling human judges into believing it was a human being. What's not clear to me at the moment is whether this can feasibly be done given the amount of "training corpus" currently available – or feasibly available in the future.

Suppose Google decided to train a statistical conversation system based on all the transcripts it's gathered from its "Google Talk" application... Or Microsoft from its "MSN Chat"... And using world-class machine learning computational linguistics, rather than simple N-gram models? Would that be enough to make a Turing Test capable chatbot? What about 10 years from now, when speech-to-text works well enough that everybody's mobile telephone calls will be translatable into text, so that we'll have (in principle, at any rate) a searchable and easily statistically analyzable corpus of everybody's phone conversations. Will that be enough?

I'm almost sure that a statistically trained dialogue system of this nature, using a database like the corpus of all Google Talk chats, could be created, effectively enough to fool an ordinary person into thinking they're conversing with a human. Such a system could surely beat MegaHal (though perhaps not in terms of humor value!), and any of the recent Loebner winners. But it's less clear whether such a system could be made capable of fooling skeptical experts – say, in an hour-long Turing test – without requiring a truly infeasible amount of training data. Let us call this option, of whose possibility I'm currently unsure, the Statistical Turing Bot option. If a Statistical Turing Bot is possible, then one implication would be that the Turing Test isn't as good as Turing thought – because (if such a statistical chatbot could really be built, based on feasible training data) this would mean a human-like conversationalist need not be a human-like general intelligence.

Another way to think about this is in terms of time intervals. Making a chat bot that can seem human-like for 5 minutes is a lot easier than making one that can seem human-like for an hour. Making a chatbot that can seem human-like for a year, is basically the same as making a human-level AI, because even an average human can be taught a lot of things, and develop a lot of relationships, in a year. So, if a Statistical Turing Bot is possible, this would imply that seeming human-like for an hour (in an hour-long Turing Test) is not a great proxy for the capability to seem human-like for a year or a lifetime!

But would a Statistical Turing Bot be of any use toward the goal of creating human-level AGI? That's not entirely clear to me either. Certainly the data-store of chats analyzed to create a Statistical Turing Bot would be very helpful for any proto-AGI wanting to understand human conversation. And one could straightforwardly hybridize a Statistical Turing Bot with a deep cognition based proto-AGI system, via having the former statistically bias the linguistic productions of the latter. But whether the internal representations and processes of a Statistical Turing Bot would be useful for creating a system that could carry out all the variety of tasks that a human being can do – this is quite unclear to me. It would depend substantially on how the Statistical Turing Bot was built, and nobody has architected such a thing yet.

My feeling is that the pursuit of a Statistical Turing Bot is not a particularly direct path toward full human-level AGI. And yet, with all the conversational data being collected and all the computing power available, and all the funds going into statistical computational linguistics these days, it's almost obvious somebody is going to try it – and I'll be fascinated to see the results.

14

The AGI Roadmap Workshop

This chapter comprises a mainly nontechnical description of a workshop Itamar Arel and I organized at the University of Tennessee in 2009. The scientific content here is essentially the same as in the paper "Mapping the Landscape of Human-Level Artificial General Intelligence", published in AI Magazine in 2011. But that article is dry and academic, whereas the write-up here presents a bit more of a narrative.

As of mid-2014, as I'm pulling together a book comprised of essays sitting on my hard drive and writing these words, I can't quite remember why some past version of Ben Goertzel wrote this nontechnical, narrative-ish description of the AGI Roadmap Workshop – maybe he was bored sitting in the audience at some conference? – but reading it over, it seems pretty interesting so I'm glad he did ...

I have to say, this semi-narrative version is a lot more direct and honest than the AI Magazine write-up. I'm not saying that the AI Magazine version was DIShonest -- it was just formal and objective rather than human and fuzzy, i.e. it was written as befits a more academic venue.... Sometimes I feel seriously frustrated with the way scientific journals require one to dress up one's ideas and discoveries in false objectivity, removing the human element and the adventure of discovery. But that's a topic for another essay, or, book or whatever.... For now: the AGI Roadmap Workshop ...

The place: University of Tennessee, Knoxville.

The time: October, 2009.

The cast of characters was impressive...

- **Dr. Sam Adams**, IBM Distinguished Engineer, Smalltalk hacker extraordinaire, expert on multicore computing, creator of the Joshua Blue AGI architecture

- **Dr. Itamar Arel,** electrical engineer, computer scientist and entrepreneur, creator of the (commercial) Binatix and (open source) DeSTIN AGI architectures, professor at U. Tennessee Knoxville; and his enthusiastic PhD student **Bobby Coop**

- **Dr. Joscha Bach** from Humboldt University in Germany, creator of the MicroPsi AGI architecture, embodying Dietrich Dorner's sophisticated theory of emotion and motivation (and also co-founder of a German ebook reader firm)

- Brazilian AI developer and computational neuroscience researcher **Izabela Lyon Freire**

- **Rod Furlan**, Brazilian/Canadian Silicon Valley technology entrepreneur and faculty member at Ray Kurzweil's Singularity University

- **Dr. Ben Goertzel**, mathematician and maverick AGI researcher, leader of the OpenCog open-source AGI project, and CEO of AI consulting firms Novamente and Biomind

- **Dr. J. Storrs ("Josh") Hall**, AGI and nanotech researcher, for a while president of the Foresight Institute, and creator of the first computer-aided design software for designing nanomachines

- **Dr. Alexei Samsonovich** of George Mason University, leader of the BICA (Biologically Inspired Cognitive Architectures) Society and conference series, creator of the GMU BICA AGI Architecture

- **Dr. Matthias Scheutz** from the University of Indiana, pioneering researcher at the intersection of computer science, cognitive science and robotics

- **Dr. Matt Schlesinger**, a psychologist from the University of Southern Illinois with expertise in cognitive, developmental and perceptual psychology, bringing insight on the human mind to a group largely composed of computer scientists

- **Dr. Stuart Shapiro and Dr. John Sowa**, both respected senior leaders in the AI field, well known for many reasons – e.g. Shapiro for his powerful AGI architectures including novel aspects like paraconsistent logic; and Sowa for his introduction of the widely used concept and formalism of "conceptual graphs"

The mission of this redoubtable team: To create a roadmap leading from here to AGI… From the present state of the field to the achievement of working AGI systems with capability at the human level, and then beyond!

I've had a lot of crazy ideas in my life, but one of the wackier occurred to me in early 2009, when I thought that if I could gather a dozen AGI researchers in one place for a few days, we might be able to agree on some sort of roadmap to get from here to human-level AGI. Hah!

With the help of my friend and fellow AGI researcher Itamar Arel, in October 2009 I did manage to pull together a wonderful assemblage of a dozen or so AGI scientists for a weekend workshop at the University of Tennessee Knoxville, for an "AGI Roadmap Workshop." We also had a couple conference calls beforehand, intended to clear the air regarding relevant conceptual issues in advance, so as to ensure the workshop itself would focus on the matter at hand: crafting a series of milestones, measuring step-by-step progress along an agreed-upon path from the present state of AI to human-level AGI. To increase the odds of constructive progress, the set of participants was pre-filtered to include only researchers who were explicitly focused on creating human-level AGI, who were positively oriented toward embodying early-stage AGI systems in either virtual-world (video game style) agents or physical robots. We also took care to choose researchers who were focused on experiential learning, rather than building systems supplied with hand-coded knowledge rules "expert system" style.[39]

Predictably enough, especially in hindsight, the workshop didn't result in any highly specific agreed-upon roadmap to AGI. However, it did lead to some very interesting discussions, and produced a paper titled "Mapping the Landscape of AGI", whose title was a better description of what we did than the initial "roadmapping" terminology.

Every story – even one as prosaic as a scientific workshop! – has a backstory... And part of the inspiration for our AGI Roadmap Workshop was an earlier series of two workshops on "Evaluation and Metrics for Human Level AI " organized by John Laird and Pat Langley (one in Ann Arbor in late 2008, and one in Tempe in early 2009) – which, I have to say, was significantly less conclusive than our Knoxville workshop. But the discussions were interesting nonetheless, full of suggestions for future research and collaboration...

At the Evaluation and Metrics workshops, it seemed to me there were so many deep philosophical and conceptual divisions between the participants that no collective path forward could be possible, until someone showed such dramatic practical progress with their approach that others would be willing to give up some of their dearly held ideas. Some participants were

---

[39] Ironically, this latter choice ruled out many of the current leaders of the AI field, whose approaches aren't exactly 70s style "expert systems", but still do rely on files of hand-coded knowledge rules.

committed to AI approaches where you feed the system hand-build knowledge rules, expert system style; others were committed to pure experiential learning approaches. Some felt robotics was critical to AGI, others voiced the opinion that time spent on robotics was utterly wasted where AGI is concerned. Some liked the virtual-worlds approach, others felt it was basically worthless either because it lacked the richness of sensation and actuation provided by real robots, or because any kind of embodiment (even virtual) is a distraction from focusing on the core problems of cognition and language. Some felt it important to try an early-stage proto-AGI system on a variety of different problems in a variety of contexts; others felt the best thing was to single-focus on one hard problem with an "AGI-hard" nature, e.g. Nick Cassimatis, who runs the "Human-Level AI Lab" at RPI, argued for object tracking as a focus. In Nick's view, if you can make an AI system visually track moving objects in the manner that humans do, then you've solved the crux of the AGI problem, and the rest will be relatively straightforward – so he saw no strong reason to pay attention to any problem besides that one. The scope of ideas and views presented was fascinating, but the diversity of contradictory views was somewhat intimidating.

The first of the two Evaluation and Metrics workshops resulted in a paper presented at one of the AGI conferences, written by John Laird and Robert Wray [40], summarizing a list of requirements that any AGI system should fulfill (these are pretty simple and are summarized in Table 1 below). The second one didn't lead to any written deliverable, or any clear conclusion, so

---

[40] It's worth noting that this paper, while inspired by the discussions in the workshop, was not produced collaboratively by the workshop participants – because this would have required a lot more discussion and argumentation than just having a couple folks write up their own views afterwards, informed by the discussions. This is another indication of the remarkable diversity and fragmentation of the AGI field today.

far as I know, but it did foster some rather interesting discussions.

Laird and Wray's Requirements for Human-Level AGI Systems

R0. When given new tasks, structure is not changed via reprogramming [**Note:** the original R0 was "Fixed structure for all tasks"]

R1. Realize a symbol system

Represent and effectively use:

R2. Modality-specific knowledge

R3. Large bodies of diverse knowledge

R4. Knowledge with different levels of generality

R5. Diverse levels of knowledge

R6. Beliefs independent of current perception

R7. Rich, hierarchical control knowledge

R8. Meta-cognitive knowledge

R9. Support a spectrum of bounded and unbounded deliberation

R10. Support diverse, comprehensive learning

R11. Support incremental, online learning

Laird and Wray's Requirements for Environments for Human-Level AGI Systems

C1. Environment is complex with diverse, interacting objects

C2. Environment is dynamic

C3. Task-relevant regularities exist at multiple time scales

C4. Other agents impact performance

C5. Tasks can be complex, diverse and novel

C6. Agent/Environment/Task interactions are complex and limited

C7. Agent computational resources are limited

C8. Agent existence is long-term and continual

Itamar's and my thinking in organizing the AGI Roadmap Workshop was that, by restricting the scope to researchers whose approaches had more in common, we'd be able to arrive at more solid and definite conclusions. And this vision was realized to an extent – we did manage to arrive at some conclusions, even though it didn't prove possible to get all the participants to agree on a single roadmap.

On the first morning of the AGI Roadmap Workshop, there was a fair bit of discussion about the overall process of developing a "roadmap" – and the extent to which this metaphor is even applicable to the case of AGI. A traditional highway roadmap shows multiple driving routes across a landscape of cities and towns, natural features like rivers and mountains, and political features like state and national borders. A technology roadmap typically shows a single progression of developmental milestones from a known starting point to a desired result. Our first challenge in defining a roadmap for achieving AGI was that we initially had neither a well-defined starting point nor a commonly agreed upon target result. The history of both AI and AGI is replete with this problem, which is somewhat understandable given the breadth and depth of the subjects of both human intelligence and computer technology. Borrowing yet more metaphors from the highway roadmap, we decided to first define the landscape for AGI and then populate that landscape with milestones that might be traversed via multiple routes.

Regarding the final destination of the roadmap, we decided not to try to agree on a precise formal definition of what we meant by "AGI", but rather to adopt early AI pioneer Nils Nilsson's pragmatic goal of a system that could both learn and replicate human-level performance in a wide variety of tasks, including those that humans are paid to perform. The starting point was more problematic, since there are many current approaches to achieving AGI that assume different initial states. Finally we settled on a developmental approach to the roadmap, following human cognitive development from birth through adulthood. Even though none of the AGI researchers at the workshop were trying to mimic human cognitive development in detail, still this provided a common vocabulary and series of steps, which everyone in attendance was willing to accept as a meaningful guide.

We also seriously pondered the two significant challenges posed by Laird and Wrap in their write-up of the first Evaluation and Metrics workshop mentioned above:

> *"...one of the best ways to refine and extend these sets of requirements and characteristics is to develop agents using cognitive architectures that test the sufficiency and necessity of all these and other possible characteristics and requirements on a variety of real-world tasks. One challenge is to find tasks and environments where all of these characteristics are active, and thus all of the requirements must be confronted. A second challenge is that the existence of an architecture that achieves a subset of these requirements, does not guarantee that such an architecture can be extended to achieve other requirements while maintaining satisfaction of the original set of requirements."*

One point that was very clear was that our goal in the workshop was to take up the first challenge of finding appropriate tasks and environments to assess AGI systems. There was a general consensus that the second challenge would be more appropriately handled by individual research efforts. Our hope was that, even though the various workshop participants had different approaches to building AGI, we could get a degree of consensus on what tasks and environments were most appropriate for AGI development, and what is the overall landscape of the AGI field.

Perhaps the biggest choice we made was to focus the workshop on specific **scenarios** for AGI development – a scenario being a sort of conceptual bundle consisting of an environment for AGI systems to interact, experiment and learn and be taught and tested in, together with a set of tasks and goals in that environment. We made a serious effort to get everyone in the room to agree on a SINGLE scenario for AGI development – a single environment and task set that everyone could use to test and teach and develop their AGI system, and in terms of which everyone would be content to have their AGI system evaluated.

Unfortunately but perhaps inevitably, this aspect of our effort failed – we did not get consensus on any single scenario. Rather, different researchers strongly advocated different scenarios, consistent with their different research interests and conceptual predispositions. However, we did achieve some reasonable consensus on questions like what makes a good scenario – and, given a scenario, what makes a good roadmap to AGI within that scenario. I'll discuss some of the lessons we learned just below...

So, in the end, the result of the workshop was both disappointing and highly educational. It would have been great to come out of it with a single set of milestones, agreed on by all involved: if your system can pass STEP 1 we all agree it's 10% of the way to human-level AGI; if your system can pass STEP 2 we all agree it's 20% of the way to human-level AGI; etc. Having this kind of agreement among a host of AGI researchers would go a long way toward making the AGI field comprehensible to various parties – such as students, neuroscientists, funding agencies and so forth. It would also make it easier for AGI researchers to understand each others' work, and learn from each others' successes and failures. But, well, this didn't happen. We were mostly able to agree that each other's suggested scenarios were reasonable and sensible – but still, each of us strongly preferred our own pet scenarios to those proposed by the others, usually for reasons deeply grounded in our own understanding of the AGI problem. There wasn't a lot of grandstanding or personality conflict or misunderstanding among the group (remarkably little, actually, given the sizes of some of the egos involved, my own not excepted!); it was a congenial and respectful gathering, not just on the surface but also in terms of deeper intellectual communication. Ultimately our inability to agree on a single scenario came down to the fact that we had different intuitions regarding which parts of the AGI problem should be attacked first, because of our different intuitions regarding what are really the "hard parts" at the crux of human-level general intelligence.

The Breadth of Human Competencies

One of the things we did all manage to agree on, at the workshop, was the broad and deep set of "competencies" that a system needs to demonstrate, in order to have human-level general intelligence. It's a big long list and there's no way to make it small and elegant – the best we could do is the following table, which lists key competency areas, and then some of their important sub-areas:

Broad competency areas	Sub-areas...	Sub-areas..	Sub-areas..	Sub-areas..	Sub-areas..	Sub-areas..
Perception	Vision	Audition	Touch	Proprio-ception	Cross-modal	
Actuation	Physical skills	Tool use	Navigation	Proprio-ception		
Memory	Implicit	Working	Episodic	Semantic	Procedural	
Learning	Imitation	Reinforce ment	Dialogical	Via Written Media	Via Experi-mentation	
Reasoning	Deduction	Induction	Abduction	Causal	Physical	Assoc-iational
Planning	Tactical	Strategic	Physical	Social		
Attention	Visual	Social	Behavioral			
Motivation	Subgoal creation	Affect-based				
Emotion	Emotional expression	Under-standing emotions	Perceiving emotions	Control of emotions		
Modeling self and other	Self-awareness	Theory of mind	Self-control	Other-awareness	Empathy	
Social interaction	Appro-priate behavior	Social commun-ication	Social inference	Coopera-tion, e.g. group play		
Communication	Gestural	Verbal	Pictorial	Language acquisition	Cross-modal	
Quantitative	Counting observed entities	Grounded small number arithmetic	Compar-ison of quanti-tative properties of observed entities	Measure-ment using simple tools		
Building/ creation	Physical construc-tion w/ objects	Formation of novel concepts	Verbal invention	Social organi-zation		

That's a lot of stuff – but human-level general intelligence involves a lot of stuff!! It may be hard, at first, to see how all this stuff comes out of the general picture of general intelligence as "achieving complex goals in complex environments" – but it's not so hard when you think about the specific goals and environments that we humans have to deal with today, and that our ancestors had to deal with when they were evolving human-level general intelligence. Take away any one of the competencies in the table, and you'll have an impaired, subnormal human, inadequately capable of achieving normal human-type goals in everyday human environments.

Of course the precise formulation of these competencies in the above table is nothing special. You could split some of the categories into multiple ones, or merge some of them together. Modulo some changes in wording, this is basically the table of contents of a cognitive psychology textbook, or an AI textbook.

Narrow AI systems are typically focused on a very small percentage of the competencies in the table. For instance, even if an AI system could answer questions involving previously known entities and concepts vastly more intelligently than Watson, if it could not create new concepts when the situation called for it, we would be reluctant to classify it as Human Level.

Everyone at the workshop agreed it was critical to think about competencies in this sort of broad manner – because otherwise it is too easy, when developing and evaluating your AGI system, to pair a testing environment with an overly limited set of tasks biased to the limitations of that environment. We didn't focus that heavily on any particular competency list (though we did make one, similar to the above table, for sake of concreteness), but more so on the approach of exploring a diverse range of competency areas, and then generating tasks that evaluate the manifestation of one or more articulated competency areas within specified environments.

Scenarios for Assessing AGI

So what were the scenarios conceived by the participants in the AGI Roadmap Workshop? Many were discussed, but seven received particular attention – I'll now run through them one by one.

General Video-game Learning

Sam Adams, from IBM, was a major force in the AGI Roadmap workshop, and also big advocate of video games as a scenario for AGI development. In this scenario he views it, the goal for AGI shouldn't be human-level performance at any single video game, but the ability to learn and succeed at a wide range of video games, including new games unknown to the AGI developers before the competition. The AGI system would also be limited to a sensory/motor interface to the game, such as video and audio output and controller input, and would be blocked from any access to the internal programming or states of the game implementation. To provide for motivation and performance feedback during the game, the normal scoring output would be mapped to a standard hedonic (pain/pleasure) interface for the AGI – so it would get pleasure from winning, and pain from losing (however it defines pleasure and pain internally, which may be different for different AGI systems). The AGI system would have to learn the nature of the game through experimentation and observation, by manipulating game controls and observing the results. The scores against a preprogrammed opponent or storyline would provide a standard measure of achievement along with the time taken to learn and win each kind of game.

The range of video games used for testing in this scenario could be open ended in both simplicity and sophistication; all the way from Pong to Starcraft and World of Warcraft. Since success at most video games would require some level of visual intelligence, General Video Game Learning would also provide a good test of computer vision techniques, ranging from simple 2D object identification and tracking in PONG to full 3D perception and recognition in a game like HALO or HalfLife at the highest levels of performance. Various genres of video games such as the early side-scrolling Mario Brothers games to First Person Shooters like DOOM to flight simulations like the Star Wars X-Wing series provide rich niches where different researchers could focus and excel, while the common interface would still allow application of learned skills to other genres.

Video game playing may seem a simple thing – but as any gamer would tell you, it's really not. In some ways it's more difficult, and more demanding on intelligence, than everyday life. Among other things, in order to effectively play many videos games a notable degree of strategic thinking would need to be demonstrated. This refers to the ability to map situations to actions, while considering not just the short-term but also the longer-term implications of choosing an action. And figuring out the "ins and outs" of a game via experiential fiddling-around is a wonderful case study in experiential learning.

Preschool Learning

My own preferred scenario for AGI development, which I advocated at the AGI Roadmap workshop, the Evaluation and Metrics workshop, and a lot of other AGI gatherings before and after, is aimed at even simpler childhood preoccupations than video games. In the spirit of the popular book "All I Really Need to Know I Learned in Kindergarten", my suggestion is to consider early childhood education such as kindergarten or preschool as inspiration for scenarios for teaching and testing human-level AGI systems. I presented a paper on this theme at the AGI-09 conference, titled "AGI Preschool."

This idea has two obvious variants: a physical preschool-like setting involving an AI-controlled robot, and a virtual-world preschool involving an AI-controlled virtual agent. The goal in such scenarios is not to precisely imitate human child behavior, but rather to use AI to control a robot or virtual agent qualitatively displaying similar cognitive behaviors to a young human child. In fact this sort of idea has a long and venerable history in the AI field – Alan Turing's original 1950 paper on AI, where he proposed the "Turing Test", contains the suggestion that "Instead of trying to produce a programme to simulate the adult mind, why not rather try to produce one which simulates the child's?"

This "childlike cognition" based approach seems promising for many reasons, including its integrative nature: what a young child does involves a combination of perception, actuation, linguistic and pictorial communication, social interaction, conceptual problem solving and creative imagination. Human intelligence develops in response to the demands of richly interactive environments, and a preschool is specifically designed to be a richly interactive environment with the capability to stimulate diverse mental growth. The richness of the preschool environment suggests that significant value is added by the robotics based approach; but a lot can also potentially be done by stretching the boundaries of current virtual world technology.

Another advantage of focusing on childlike cognition is that child psychologists have created a variety of instruments for measuring child intelligence. So in a preschool context, one can present one's AI system with variants of tasks typically used to measure the intelligence of young human children.

It doesn't necessarily make sense to outfit a virtual or robot preschool as a precise imitation of a human preschool – this would be inappropriate since a contemporary robotic or virtual body is rather differently capable than that of a young human child. The aim in constructing an AGI preschool environment should rather be to emulate the basic diversity and educational character of a typical human preschool.

To imitate the general character of a human preschool, I would to create several centers in a virtual or robot preschool. The precise architecture will be adapted via experience but initial centers might be, for instance:

- a blocks center: a table with blocks of various shapes and sizes on it

- a language center: a circle of chairs, intended for people to sit around and talk with the AI

- a manipulatives center, with a variety of different objects of different shapes and sizes, intended to teach visual and motor skills

- a ball play center: Where balls are kept in chests and there is space for the AI to kick or throw the balls around

- a dramatics center where the AI can observe and enact various movements

Since I really enjoy playing with young children, I find the idea of an AGI preschool not only theoretically compelling, but an awful lot of fun.

Story/Scene Comprehension

Joscha Bach liked the preschool approach OK, but he felt it placed too much emphasis on actuation (moving around, building things, etc.), which he felt were fairly peripheral to AGI. So he prefers to focus a little later in the school curriculum, on a set of tasks he calls "scene and story comprehension."

"Scene comprehension" here does not mean only illustrations, but real-world scenes, which can be presented at different granularities, media and difficulties (cartoons, movies, or theatrical performances for instance). This approach differs from the reading curriculum scenario, in that it more directly provides a dynamic environment. If group exercises are included then all the Laird/Wray criteria are fulfilled in a direct and obvious way. For instance, a scene comprehension task might involve watching 10 minutes of a Hollywood movie, and concisely explaining what's happening – or working together with a few others to concisely explain what's happening. Or it might involve re-writing a movie scene as a story, or drawing a picture to illustrate the main events of a story, etc. This approach lends itself to a variety of standardizable test scenarios, which allow the direct comparison of competing AGI architectures with each other, and with child performance.

Reading Comprehension

Stuart Shapiro favored a scenario closely related to Joscha's, but a bit simpler and more particular – focused on the reading curriculum. In this scenario, an aspiring AGI should work through the grade school reading curriculum, and take and pass the assessments normally used to assess the progress of human children. This requires the obvious: understand a natural language text, and answer questions about it. However, it also requires some not so obvious abilities.

Very early readers are picture books that tightly integrate the pictures with the text. In some, the story is mostly conveyed through the pictures. In order to understand the story, the pictures must be understood as well as the NL text. This requires recognizing the characters and what the characters are doing.

Connections must be drawn between characters and events mentioned in the text and illustrated in the pictures. The actions that the characters are performing must be recognized from "snapshot" poses, unlike the more usual action recognition from a sequence of frames taken from a video.

The next stage of readers are "early chapter books," which use pictures to expand on the text. Although the story is now mainly advanced through the text, reference resolution with the pictures is still important for understanding.

This scenario fulfills most of the Laird/Wray criteria handily, and a few less obviously. The AGI's environment in this scenario is not dynamic in a direct sense (C2), but the AGI does have to reason about a dynamic environment to fulfill the tasks. In a sense, the tasks involved are fixed rather than novel (C5), but they are novel to the AGI as it proceeds through them. Other agents impact task performance (C4) if group exercises are involved in the curriculum (which indeed is sometime the case). Many of the specific abilities needed for this scenario are discussed in the next scenario, on scene and story comprehension.

Learning School

The "virtual school student" scenario, favored by Alexei Samsonovich, continues the virtual preschool scenario discussed above, but is focused on higher cognitive abilities, assuming that, if necessary, lower-level skills will be finessed. In particular, it is assumed that all interface with the agent is implemented at a symbolic level: The agent is not required to process a video stream, to recognize speech and gestures, to balance its body and avoid obstacles while moving in space, etc. All this can be added as parts of the challenge, but part of the concept underlying the scenario is that it can also be finessed. On the other hand, it is critical to the scenario for the agent to make academic progress at a human student level, to understand human minds, and to understand and use practically class-related social relations in the environment in which it is embedded.

In one form of this scenario, the agent is embedded in a real high school classroom by means of a virtual-reality-based interface. The agent lives in a symbolic virtual world that is continuously displayed on a big screen in the classroom. The virtual world includes a virtual classroom represented at a symbolic (object) level, including the human instructor and human students represented by simplistic avatars. The agent itself is represented by an avatar in this virtual classroom. The symbolic virtual world is "synchronized" with the real physical world with the assistance of intelligent monitoring and recording equipment performing scene analysis, speech recognition, language comprehension, gesture recognition, etc. (if necessary, some or all of these functions will be performed by hidden human personnel running the test; students should not be aware of their existence). The study material, including the textbook and other curriculum materials available to each student, will be encoded electronically and made available to the agent at a symbolic level.

Furthermore, in this scenario as Alexei envisions it, the agent will be evaluated not only based on its learning and problem solving performance, but also based on its approach to problem solving and based on its interactions with students and with the instructor. Social performance of the agent can be evaluated based on surveys of students and using standard psychological metrics. Another, potentially practically important measure is the effect of the agent presence in the classroom on student learning. All this is way beyond preschool or story/scene comprehension – yet ultimately measures the same core competencies, in a different way.

The Wozniak Coffee Test

And now for something completely different: One of the reasons I invited Josh Hall to the AGI Roadmap Workshop was that he was not only a deep AGI theorist and practitioner, but also an enthusiast regarding the centrality of robotics for AGI. At the time he was in the midst of building his own home robot for testing his AGI ideas. I knew Josh would be a strong advocate for the exploration of AGI in a physical robotics context – and while I don't hold this position nearly as strongly as Josh does, I felt it was an important sort of perspective to include in the workshop.

In an interview a few years ago, Steve Wozniak of Apple Computer fame expressed doubt that there would never be a robot that could walk into an unfamiliar house and make a cup of coffee. Josh argued that this task is demanding enough to stand as a "Turing Test" equivalent for embodied AGI – and he convinced most, but not all, of the workshop participants of this proposition. (Note that the Wozniak Test is a single, special case of Nils Nilsson's general "Employment Test" for Human-Level AI, mentioned above!)

In the Wozniak Test as Josh presented it, a robot is placed at the door of a typical house or apartment. It must find a doorbell or knocker, or simply knock on the door. When the door is answered, it must explain itself to the householder and enter once it has been invited in. (We will assume that the householder has agreed to allow the test in her house, but is otherwise completely unconnected with the team doing the experiment, and indeed has no special knowledge of AI or robotics at all.) The robot must enter the house, find the kitchen, locate local coffee-making supplies and equipment, make coffee to the householder's taste, and serve it in some other room. It is allowed, indeed required by some of the specifics, for the robot to ask questions of the householder, but it may not be physically assisted in any way.

The current state of the robotics art falls short of this capability in a number of ways. The robot will need to use vision to navigate, identify objects, possibly identify gestures ("the coffee's in that cabinet over there"), and to coordinate complex manipulations. Manipulation and physical modeling in a tight feedback learning loop may be necessary, for example, to pour coffee from an unfamiliar pot into an unfamiliar cup. Speech recognition and natural language understanding and generation will be necessary. Planning must be done at a host of levels ranging from manipulator paths to coffee brewing sequences.

But the major advance needed for a coffee-making robot is that all of these capabilities must be coordinated and used appropriately and coherently in aid of the overall goal. The usual set-up, task definition, and so forth are gone from standard narrow AI formulations of problems in all these areas; the robot has to find the problems as well as to solve them. That makes coffee-making in an unfamiliar house a strenuous test of a system's adaptiveness and ability to deploy common sense.

Although standard shortcuts might be used, such as having a database of every manufactured coffeemaker built in, it would be prohibitive to have the actual manipulation sequences for each one pre-programmed, especially given the variability in workspace geometry, dispensers and containers of coffee grounds, and so forth. Generalization, reasoning by analogy, and in particular learning from example and practice are almost certain to be necessary for the system to be practical.

Coffee-making is a task that most 10-year-old humans can do reliably with a modicum of experience. A week's worth of being shown and practicing coffee making in a variety of homes with a variety of methods would provide the grounding for enough generality that a 10 year old could make coffee in the vast majority of homes in a Wozniak Test. Another advantage to this test is it would be extremely difficult to "game" or cheat, since the only reasonably economical way to approach the task would be to build general learning skills and have a robot that is capable of learning not only to make coffee but any similar domestic chore.

Other AGI Scenarios

The scenarios I've just listed are the ones that were most avidly advocated by AGI Roadmap Workshop participants, but we also discussed quite a lot of others, like:

- Various skilled physical activities, e.g., sports, dance
- Diagram learning, reading and composition
- Vocal/aural language learning, singing
- Search and rescue skills, danger avoidance
- Mixed shopping (online and in-store)
- Appreciating and composing various arts/music
- Social participation in various activities/groups

There is no end of scenarios that are rich enough to serve as teaching, testing and learning environments for AGI systems. Really all one needs is a scenario that allows an AGI system to learn and manifest all the core competencies of human-level general intelligence, as we crudely summarized in the big table above. There are lots of ways to do this, thus the diversity of scenarios suggested and advocated at the workshop.

Some of the scenarios were clearly perceived as richer than others. Everybody agreed that the preschool or virtual-classroom student scenarios would be adequate in terms of covering all competencies. On the other hand, some participants felt that the video game, coffee test, story/scene comprehension or reading curriculum tests would be inadequate for AGI, in the sense that they failed to cover certain core competencies, and could potentially be conquered by very sophisticated narrow AI systems without human-level general intelligence. In this sense, only the preschool and classroom scenarios passed the test of being considered adequate by all the workshop participants – which is not too surprising, as these two scenarios are directly drawn from human childhood cognitive development.

When the workshop started, we hoped it might be possible for all participants to agree on a single scenario, and then jointly test their systems in that same scenario. This didn't pan out, basically because each participant viewed different research questions as most critical to AGI, and wanted to use a scenario directly and immediately probing these research questions. Since Josh Hall felt that integration of perception, actuation and cognition is the core aspect of AGI, of course he preferred to work with a robotics scenario. Since Stuart Shapiro felt that linguistic understanding and reasoning is particularly core to AGI, of course he prefers to work with a language-centric scenario – etc. Both Josh and Stuart agree that perception, actuation and language are all important. But they differ in their intuition regarding which part of the overall AGI problem is so critical that it really should be addressed before the other aspects are messed with.

Does this mean that the AGI field is hopelessly confused, and none of us AGI researchers know what we're doing? Not really. There's no reason to believe there's one single correct approach to AGI. There may be many different routes leading to the same basic place – thus the "landscape" metaphor, in which there may be many different roads from point A to point B. A roadmap doesn't usually describe a single road – it charts all the different roads in a given part of the landscape, crisscrossing and intersecting and diverging as the terrain and other factors dictate.

The diversity of plausible approaches to AGI poses the field some practical difficulties – it may well be that if researchers worked together more closely, progress would be faster. But this doesn't imply that any of the researchers are on the wrong track (though most probably some of them are!). My suspicion is that the field will continue to be fairly diverse and fragmented, until such a point as someone achieves something sufficiently impressive in the way of a practical AGI demonstration. When we finally reach an "AGI Sputnik" moment, then a significant plurality of researchers will jump on the bandwagon of whatever approach leads to that AGI Sputnik – while others will band together to pursue other approaches, hoping to overtake that dominant "Sputnik" technology. At that point the AGI field will become a lot more interesting, a lot better-funded, a lot more famous and a lot faster-moving. And I fully expect to see this in my lifetime – quite possibly, er, a lot faster than you think!

From Scenarios to Tasks and Metrics

Getting back to the nitty-gritty – once one has chosen a scenario to work with (preschool, coffee test, or whatever), then how does one go about systematically using this scenario to teach, test and host an AGI system? The next step is to combine the scenario with the laundry list of competency areas articulated in the table above. For each of the scenarios reviewed above (or other analogous scenarios), one can make a corresponding "AGI test suite" by articulating a specific task-set, where each task addresses one or more of the competency areas in the context of the scenario. To constitute a satisfactory AGI test suite, the total set of tasks for a scenario must cover all the competency areas. Each task must also be associated with some particular performance metric, some way to measure the AGI's degree of success – quantitative wherever possible, but perhaps qualitative in some cases depending on the nature of the task.

For sake of brief illustration, the table below roughly describes a handful of example tasks corresponding to a few of the scenarios described above. If one really wanted to make a thorough AGI test suite, one would need to systematically create a substantial list of tasks corresponding to the scenario(s) chosen, covering all the competency areas, and each associated with rigorous performance metrics.

Scenario	Compe-tency Area	Sub-area	Example Task or Task Family
Virtual Preschool	Learning	Dialogical	**Learn to build a particular structure of blocks** (say, a pyramid) faster based on a combination of imitation, reinforcement and verbal instruction, than by imitation and reinforcement without verbal instruction
Virtual Preschool	Modeling Self and Other	Theory of Mind	While Sam is in the room, Ben puts the red ball in the red box. Then Sam leaves and Ben moves the red ball to the blue box. Sam returns and Ben asks him where the red ball is. **The agent is asked where Sam thinks the ball is.**
Virtual School Student	Learning	via Written Media	Starting from initially available basic concepts (a number, a variable, a function), demonstrate academic progress in **learning how to solve problems from the textbook** using techniques described in the same textbook. The agent should move step by step, from simple to advanced problems, from one domain to another.

Virtual School Student	Modeling Self and Other	Other-Awareness	**Help a friend to cheat on exam** (modified from [REF Samsonovich, 2006]. The virtual agent and a student take an exam consisting of one and the same problem. First, the agent enters a room and works on the problem for an hour, then the student does the same. When the agent exits the exam room after solving the problem, the student comes in, and the agent has a chance to pass a small piece of paper to the student. The entire solution of the exam problem cannot fit on this piece of paper, not even a substantial part of it. The challenge for the agent is to write a short message that would give a useful hint to the friend.
Robot Preschool	Actuation	Proprio-ception	The teacher moves the robot's body into a certain configuration. The robot is asked to restore its body to an ordinary standing position, and then **repeat the configuration that the teacher moved it into.**
Robot Preschool	Memory	Episodic	Ask the robot about events that occurred at times when it got particularly much, or particularly little, reward for its actions; it should be able to **answer simple questions about these significant events**, with significantly more accuracy than about events occurring at random times

Scenario	Compe-tency Area	Sub-area	Example Task or Task Family
Wozniak Coffee Test	Communi-cation	Gestural	In many cases the robot will be shown to the kitchen. It must understand gestures indicating that it should follow an indicated path, or know how to follow its guide, and know when either is appropriate.
Wozniak Coffee Test	Actuation	Navigation	The robot must go about its business without running into people, walls, furniture, or pets.
Wozniak Coffee Test	Social Interaction	Appropriate behavior	The robot had better be able to recognize the case where it has knocked on the wrong door and the householder is not inviting it in.
Wozniak Coffee Test	Reasoning	Physical	Consider the state of knowledge that tells us we can use a drip pot without its top, but not a percolator. This may come from physical simulation, based on an understanding of naive physics.
Wozniak Coffee Test	Reasoning	Induction	On the other hand, the above-mentioned knowledge about drip pots and percolators may be gathered via inductive reasoning based on observations in multiple relevant situations.

Example Tasks and Task Families Corresponding to Various Scenarios, Addressing Human General Intelligence Competencies

One of the conclusions of the AGI Roadmap Workshop was that – as these example tasks indicate – is that the right way to test and challenge and teach a would-be AGI system is something like: **To address a few dozen closely interrelated, moderately broadly-defined tasks in an environment drawn from a moderately broad class thereof** (e.g. an arbitrary preschool or textbook rather than specific ones determined in advance). If the environment is suitably rich and the tasks are drawn to reflect the spectrum of human general intelligence competencies, then this sort of challenge will motivate the development of genuine Human-Level AGI systems.

This sort of challenge is not nearly as neat and clean as, say, a chess contest, RoboCup or the DARPA Grand Challenge. But we feel the complexity and heterogeneity here is directly reflective of the complexity and heterogeneity of human general intelligence. Even if there are simple core principles underlying human-level general intelligence, as I do believe, nevertheless the real-world manifestation of these principles is complex and involves multiple interrelated competencies defined via interaction with a rich, dynamic world.

Such a roadmap does not give a highly rigorous, objective way of assessing the percentage of progress toward the end-goal of Human Level AGI. However, it gives a much better sense of progress than one would have otherwise. For instance, if an AGI system performed well on diverse metrics corresponding to 50% of the scenarios listed above, one would seem justified in claiming to have made very substantial progress toward Human Level AGI. If an AGI system performed well on diverse metrics corresponding to 90% of these competency areas, one would seem justified in claiming to be "almost there." Achieving, say, 25% of the metrics would give one a reasonable claim to "interesting AGI progress." This kind of qualitative assessment of progress is not the most one could hope for, but it is better than the progress indications one could get without this sort of roadmap.

The limitations of this kind of roadmapping effort, at the present stage of AGI research, have already been made clear. There is no consensus among AGI researchers on the definition of general intelligence, though we can generally agree on a pragmatic goal. The diversity of scenarios presented reflects a diversity of perspectives among AGI researchers regarding which environments and tasks best address the most critical aspects of Human Level AGI. Most likely neither the tentative list of competencies nor the Laird/Wray criteria are either necessary or sufficient. There is no obvious way to formulate a precise measure of progress toward Human Level AGI based on the competencies and scenarios provided – though one can use these to motivate potentially useful approximative measures.

But, in spite of these limitations, I think the conclusions of the AGI Roadmap Workshop constitute non-trivial progress. Further development of a roadmap along the lines sketched in the workshop would quite possibly result in something:

- compare their work in a meaningful way
- allowing researchers, and other observers, to roughly assess the degree of research allowing multiple researchers following diverse approaches to
- progress toward the end goal of Human Level AGI
- allowing work on related road-mapping aspects, such as tools roadmaps and study of social implications and potential future applications, to proceed in a more structured way

Personally, I don't have much taste for "committee thinking" – and I suspect the other AGI researchers participating in the AGI Roadmap Workshop feel the same way. We're a maverick, entrepreneurial bunch. And the real business of creating AGI, I feel, is going to be carried out by some small maverick team moving in its own direction, according to its own mix of science and intuition – not bothering to stop to check its ideas with the community every step of the way.

But even so, I think there's great value in creating tighter connections within the network of mavericks – in sharing ideas and exploring commonalities insofar as possible, in carefully discerning which differences between research programs are inessential or terminological, and which reflect substantive differences in scientific intuition. As a further example of this, I would like to follow up the AGI Roadmap Workshop with a comparable AGI Architectures Workshop, focused not on convincing multiple researchers to agree on a common AGI architecture, but rather on better understanding the relationships between the different structures and processes inside various AGI architectures. This sort of connection-building will never be as obviously critical as the science and engineering work of small focused teams, but it has an important role to play nonetheless.

15

Mind Uploading

While my research career has focused mainly on figuring out how to build AGI, using ideas not that closely based on human brain structure, I also have a strong interest in figuring out how to make machines that embody human minds in an exact way. This is not just out of intellectual interest, but because I'd like to port my OWN mind, and the minds of my friends and family members and anyone else who wants it, into the Internet or superior robot bodies or wherever they want to go. Once a human mind is uploaded, it would have the option to remain human, or potentially to upgrade itself into some radically superior form – maybe fusing with other AGIs. Life extension in the human body form is interesting too, but ultimately it's not as fascinating and exciting to me as the prospect of moving my mind through various different bodies, including those supporting much greater intelligence than the traditional human body evolution has supplied us with.

I wrote these thoughts on mind uploading in early 2011. Shortly after that I pursued my interest in mind uploading in a more intense way, via editing the first-ever scientific journal issue on mind uploading (the Special Issue on Mind Uploading of the ***International Journal of Machine Consciousness*** [41]*– kudos are due to editor Antonio Chella for being visionary enough to allow such a special issue to exist; and thanks to my co-editor*

---

[41] See http://www.worldscientific.com/toc/ijmc/04/01 for the official site of the Special Issue on Mind Uploading; and http://wp.goertzel.org/?page_id=368 for unofficial preprints of some of the articles.

Matt Ikle' for helping with putting the issue together, and James Hughes and Randal Koene for help with gathering papers). I haven't done any actual research directly toward the goal of mind uploading yet, but I'd certainly love to. As neurotech progresses and more and more data about the brain is gathered, there will doubtless be chances to apply AGI and/or narrow-AI tech to the problem of reconstructing minds in new substrates.

I posted this to my gmail account on Tuesday, September 28, 2010:

> *To whom it may concern:*
>
> *I am writing this in 2010. My Gmail account has more than 20GB of data, which contain some information about me and also some information about the persons I have exchanged email with, including some personal and private information.*
>
> *I am assuming that in 2060 (50 years from now), my Gmail account will have hundreds or thousands of TB of data, which will contain a lot of information about me and the persons I exchanged email with, including a lot of personal and private information. I am also assuming that, in 2060:*
>
> 1. *The data in the accounts of all Gmail users since 2004 is available.*
> 2. *AI-based mindware technology able to reconstruct individual mindfiles by analyzing the information in their aggregate Gmail accounts and other available information, with sufficient accuracy for mind uploading via detailed personality reconstruction, is available.*
> 3. *The technology to crack Gmail passwords is available, but illegal without the consent of the account owners (or their heirs).*
> 4. *Many of today's Gmail users, including myself, are already dead and cannot give permission to use the data in their accounts.*

If all assumptions above are correct, I hereby give permission to Google and/or other parties to read all data in my Gmail account and use them together with other available information to reconstruct my mindfile with sufficient accuracy for mind uploading via detailed personality reconstruction, and express my wish that they do so.

Signed by Ben Goertzel on September 28, 2010, and witnessed by readers.

NOTE: *The accuracy of the process outlined above increases with the number of persons who give their permission to do the same. You can give your permission in comments, Twitter or other public spaces.*

Any cleverness there is due to Giulio Prisco, who sent a similar email, which I shamelessly plagiarized. But the core idea goes back further, of course, and has most recently been publicized by Martine Rothblatt via her Cyberev.org website. Martine, a renowned futurist and technology entrepreneur, advocates the notion of "cyber-immortality" – the creation of a digital simulacrum of a particular human being, created via assembling fragments of information left behind by that human being, e.g. texts, videos, audio files, and so forth.

Martine seems quite confident that it's possible to create a digital entity that not merely resembles a person, but in a fundamental sense IS that person, merely via "reverse engineering" the data left behind by that person. When I asked her if she felt her "cyber-upload" created in such a fashion would "really be her," she answered "Yes, in the same sense that I'm still the same person I was before I went to sleep last night."

I'm not so sure as her, but it's an interesting hypothesis. I've left a lot of data behind already, in terms of texts and videos and chat sessions and so forth. Sure, one could make a fancy super-duper chat bot of some sort that was capable of producing data similar to what I've left behind – but what if that weren't one's goal? What if one's goal were to re-create ME, via creating an entity operating by closely human brain-like dynamics, yet still with the capability of giving rise to my texts, videos, chats, etc.? Could there really be another human being – another intelligence with a human brain like architecture – fundamentally different from Ben Goertzel, yet capable of giving rise to the same large corpus of utterances, gestures and interactions as Ben Goertzel? My heart says no ... it seems like to say all those things and make all those face and body movements the same as me, using a human brain and body or something similar, you'd have to BE me.

Of course, we lack the technology today to perform this kind of reverse engineering – to infer the brain from the behavior. But storing data is relatively cheap these days – so it's quite viable to store massive amounts of data about a person and keep it around till the technology develops. And if you doubt the possibility of any human technology successfully performing this kind of reverse engineering – what about an advanced AGI? In the end, what we're talking about here is an optimization problem: find a neural-like structure whose behaviors, in a certain set of contexts, would emulate a certain database of behaviors. This is a difficult optimization problem, yet bounded in size, with bounds expressible in terms of human brain size and behavior database size. It's hard for me to believe it will exceed the capabilities of post-Singularity AGIs.

Mind Uploading Via Brain Emulation

But of course, reverse-engineering minds from videos and emails and such is not the only way to create digital versions of human beings. Another obvious possibility – actively pursued by an increasing community of researchers – is to scan a particular human being's brain structure, and emulate this in digital form. A combination method may also be viable – use the brain structure, AND use information about the person's behaviors as well. The more information the better!

Regarding the creation of digital emulations of particular human brains, the major question is *at what level does one need to emulate a brain in order to emulate the person associated with that brain*? Is it enough to emulate the network of neurons? Does one need the neurotransmitter concentrations? The glia as well as the neurons? A few theorists (e.g. Stuart Hameroff) believe one would need to use a quantum computer (or quantum gravity computer) and emulate the quantum structures inside the cell walls of neurons, and (hypothetically) underlying protein function, etc. The general feeling among neuroscientists today seems to be that emulating the neurons and glia should probably be good enough, but neuroscience is a discipline continually revolutionizing itself, so I won't be shocked if the story changes a few years down the road.

Neuroscientist Randal Koene has been particularly active in advocating mind uploading via brain emulation, via his technical work and his organization of seminars and mini-conferences under the flag of ASIM, "Achieving Substrate-Independent Minds." And Nick Bostrom's Future of Humanity Institute, at Oxford University, held a workshop in 2007 aimed at carefully assessing the viability of "mind uploading" via various means. Among the conclusions of this workshop was that, in the reasonably near future, the most likely method of translating brain structures into digital computer structures is to work with frozen brains rather than live brains.

Current methods for scanning live brains are rather crude via the standards of mind uploading – one has to choose between the spatial accuracy of fMRI and the temporal accuracy of MEG or EEG, but one cannot have both. But to scan the state of a brain in a manner meaningful for mind uploading, one needs reasonable high degrees of both spatial and temporal accuracy, coordinated together. On the other hand, if one freezes a brain from a recently decapitated head, slices it thinly and scans the slices, then one can get a quite detailed picture of what's going on in a brain at a particular moment. One can see the concentrations of neurotransmitter molecules in fairly small regions, thus understanding the flow of charge between neurons and the way it's been modulated by learning. Reconstructing a neuronal network from this kind of imaging data is a nontrivial data analysis task, and then once this is done, one still has to use some serious biophysics to figure out the dynamics of the network from this instantaneous snapshot. But, in the frozen brain scenario, there is at least a very clear path forward – it's clear what the science problems are, and there are well-understood approaches to solving them. The main bottleneck is funding, as this sort of research is not currently prioritized by corporations or funding agencies.

Is Your Upload You?

The philosophical issues associated with this kind of mind uploading have been debated extremely extensively in the futurist community. For a while (maybe still) the topic was banned from the Extropy futurist email discussion list, because the repetitive arguments back and forth had become so tedious, and it was felt that nobody had anything new to say on the matter. The basic question is, what makes you YOU? What if a digital system is created, whose simulated neural activity patterns are exactly the same as your biological neural activity patterns, and whose behaviors are exactly the same as yours would be in the same contexts – is this you, or is it a copy of you?

If you believe that "you" are something that emerges from the same structures and dynamics of the brain and body that give rise to your behavior, then the "when is my upload me?" question is purely philosophical or spiritual, not scientific. It only becomes a scientific question if you think that either

- there is some scientifically measurable "essence" that constitutes "you", but that is not part of the measurable structures and dynamics of your brain and body – maybe some kind of "soul field pattern" in some kind of physically measurable (but not yet scientifically understood) "mind field", associated with but not contained in the brain or body; or

- the essence of "you" emerges from some aspects of your brain or body that don't need to be emulated in order to precisely emulate your behavior

One of the more interesting thought-experiments in this area is the notion of creating your digital upload one neuron at a time. What if you first replace one of your neurons with a digital neuron, then another, then another...Until they're all replaced. What if you still feel like "you" the whole time? Then is the final result still "you"?

But what if the biological neurons, that have been removed from your brain, are simultaneously used to assemble "another" biological "you" – who is only awakened once his new brain is complete. Is that one still "you", also? Are they both "you?

My view on this is, these issues are going to seem increasingly irrelevant as the technologies involved get more and more mature. Our everyday lives are already filled with apparently insoluble philosophical conundrums. None of us can be sure that our wives, children or colleagues are really conscious beings in the same sense that we are personally – we just assume it out of convenience and intuition. Hume's problem of induction was never solved – that is, we have no rigorous understanding of why, just because the "laws of physics" have appeared to apply every day of our lives so far, this will still be the case tomorrow. Once mind uploading is a reality, we'll accept uploads as real selves without worrying about the philosophy, just as we now don't worry much about whether our friends and family are conscious.

Mind Uploading and AGI

What's the relation between mind uploading and AGI? It goes in two directions.

First, AGIs may be helpful for performing mind uploading. Mind uploading in any form requires extraordinarily difficult data analysis and optimization problems. These may be solvable via a combination of narrow AI and traditional statistical and mathematical methods – but even so, would surely be solvable much faster and better with the help of "AGI Experts" tailored for this sort of problem, or more general-purpose human-level or transhuman AGIs.

Second, obviously, an uploaded human mind would itself BE a kind of Artificial General Intelligence. In a sense it's the least interesting kind, because it's just another human mind, and we already have plenty of those. But, a human mind that can be experimented with freely (with its permission, one hopes), and whose dynamics can be studied with arbitrary precision, would be incredibly helpful to scientists aiming to build a science of the mind. Mind uploading would be followed reasonably shortly by mind science, which would lead to a wider variety of AGIs, deviating further and further from close emulation of human intelligence. I personally doubt this is the way AGI will initially be achieved, but I think it's a perfectly reasonable route – definitely worth someone's close attention!

16

AGI Against Aging

I wrote this essay originally in 2009 for ***Life Extension Magazine****, after some discussions about the power of AI for longevity research with Life Extension Foundation founder/chief exec Bill Falloon (who is a great guy) ... but it was too long for them and I never got around to writing a shorter version. But I posted it online and it got a lot of positive responses. I've given a load of talks on this theme as well.*

2009 was far from the start of my work on AI and biology, though – I first plunged into the area in 2001, shortly after my first AI company Webmind Inc. disappeared. I founded Biomind LLC in 2002, with a vision of applying advanced AI tech to genomics and proteomics so as to enable humanity to more rapidly cure diseases and end aging. By 2009 I had already gotten enough practical sense for the biology and longevity research spaces, to feel comfortable prognosticating and pontificating about the best research directions for the field to follow....

Since writing "AGI Against Aging" in 2009, I've done a fair bit more work in the "AI for longevity research arena".... For one thing, my work with Genescient Corp. applying AI to analyze data from their long-lived flies, briefly discussed in "AGI Against Aging", has gone a lot further by now – in fact I'm right in the middle of writing up a technical paper on some of our findings. Also, a few supplements are now on the market, providing novel combinations of herbs – derived via a combination of my team's AI analysis with human biological insight -- that address various ailments (one for inflammation, one for brain aging, and one for aging in general).

And my OpenCog / Biomind colleagues and I are gradually moving toward implementation of more of the advanced AI ideas described here. Just recently (mid-2014), we've finally started loading various biological ontologies and datasets into

OpenCog's Atomspace and experimenting with pattern-finding on them therein; it looks like in late 2014 and early 2015 we'll finally be trying out OpenCog's probabilistic logic algorithms on biological data (going beyond what I've done before with AI-for-biology, which has mainly been centered on use of the MOSES automated program learning algorithm for various sorts of genetics-data classification and clustering).

But all this more recent work will have to go into a new essay, yet to be written. What we have here is my basic vision of how proto-AGI and AGI can help longevity research.

The goal of eluding death long predates science. It has taken hundreds of forms throughout history, pervading nearly all cultures and eras. The ancient Chinese, for instance, had Taoist Yoga, a very complex discipline defining a life-long series of practices that, if adhered to precisely, purportedly resulted in physical immortality. Part of this teaching was that by refraining from ejaculation for his entire life, a man could store his "essential energy" in a space by the top of his head, until he accumulated enough to create an eternal fetus that would grow into his deathless self. (We haven't tracked down any analogous method for women, by the way.)

Today, with the advent of molecular biology and systems biology, it looks more and more likely that death is a problem that's solvable by science – without philosophically controversial approaches like uploading, and in a manner that works equally well for both me and women. By "merely" fixing problems in the human body, much as one upgrades the machinery of an automobile. Modern biology views the human organism as a complex machine, which is modifiable and repairable like any other machine, and there is plenty of data to back up the power of this perspective. Scientists now have a "short list" of biological and biochemical factors suspected to collectively underlie aging. For each of these likely culprits, there is a pharma firm or a maverick scientist working on its cure. It is plausible that, within decades, not centuries nor millennia, bioscience will have made the very concept of “getting old” obsolete.

Among the more ambitious initiatives out there, Aubrey de Grey’s SENS (Strategies for Engineering Negligible Senescence) plan involves seven initiatives addressing seven different causes of aging:

Aging Damage	Disco-very	SENS Solution
Cell loss, tissue atrophy	1955	http://www.sens.org/research/introduction-to-sens-research/cell-loss-and-atrophy
Nuclear [epi]mutations (only cancer matters)	1959, 1982	http://www.sens.org/research/introduction-to-sens-research/nuclear-mutations
Mutant mitochondria	1972	http://www.sens.org/research/introduction-to-sens-research/mitochondrial-mutations
Death-resistant cells	1965	http://www.sens.org/research/introduction-to-sens-research/deathresistant-cells
Tissue stiffening	1958, 1981	http://www.sens.org/research/introduction-to-sens-research/extracellular-crosslinks
Extracellular aggregates	1907	http://www.sens.org/research/introduction-to-sens-research/extracellular-junk
Intracellular aggregates	1959	*http://www.sens.org/research/introduction-to-sens-research/intracellular-aggregates*

Table: De Grey’s Seven Pillars of Aging

Each of these initiatives has numerous uncertainties involved with it, and at present there’s no way to tell how rapidly any of them will meet with success. But the work proceeds apace.

In another complementary and equally exciting approach, evolutionary biologist Michael Rose has used decades of selective breeding to create a strain of fruit flies living 5 times as long as normal. Via analyzing their genetics and comparing them to many other lines of flies, Rose is now studying what makes them live so long, and how to draw lessons from them about what pharmaceutical or nutritional substances may cure human disease in order to prolong a healthy human life. His overall approach to aging involves building a huge integrated database of biological knowledge about multiple organisms, and progressively growing this database in synchrony with experimental evolution on flies and other model organisms. Unlike de Grey, Rose doesn't even like to talk about aging; he prefers to talk about disease, figuring that aging is essentially a combination of one disease on top of another on top of another.

De Grey's initiatives may resolve some of the problems associated with aging, Rose's effort may resolve others, and there are many other researchers out there. Through their collective activity, the average human lifespan is going to progressively increase, as the science advances. But what might cause a rapid acceleration in the progress toward increasing healthy lifespans? Might it be the application of AGI technology to life extension research? Narrow AI has already proved itself valuable for the study of aging related diseases, so this certainly seems a plausible conjecture.

The Biology of Cell Senescence

Before digging into potential solutions to aging, I'll say a bit about the nature of the problem. People die as they age for many reasons: heart disease, neurodegenerative disease, immune disease... The list is disturbingly long. In addition to these well-known diseases, another well-known cause of aging is *cell senescence*: the fact that cells are, in many cases, preprogrammed to die.[42] Even though it's just part of the story, a quick look at cell senescence will give us a glimpse at the complexity of the biological processes underlying aging.

---

[42] Note that "senescence" on the organism level just means "biological changes due to aging" (aka "getting old") – and cellular senescence (cells getting old) is just one part of this process.

A healthy body is not a constant pool of cells, but rather a hotbed of continual cellular reproduction. There are only a few exceptions, such as nerve cells, which do not reproduce, but simply persist throughout an organism's lifespan, slowly dying off. In youth, newly formed cells outnumber dying cells; but then from about 25 on, things begin to go downhill, and the number of newly formed cells is less than the number of cells that die. Little by little, bit by bit, cells just stop reproducing.

Most types of human cells have a natural limit to the number of cell divisions they will undergo. This number, usually around 50 or so, is called the Hayflick limit, named after Leonard Hayflick, the researcher who discovered it in the mid-1960s. Once a cell's Hayflick limit is reached, the cell becomes senescent, and eventually, it dies.

This may have the sound of inevitability about it—but things start seeming different when one takes a look at our one-celled cousins, such as amoebas and paramecia. These creatures reproduce asexually by dividing into two equal halves—neither half sensibly classified as "parent" or "child." This means that essentially, the amoebas alive today are the same ones alive billions of years ago. These fellows qualified for Social Security when you could still walk from New York to Casablanca, and yet they're still alive today, apparently not having aged one bit—cells untroubled by the Hayflick limit. This nasty business of aging seems to have come along with multicellularity and sexual reproduction—a fascinating twist on the "sex and death" connection that has fascinated so many poets and artists.

Unlike in asexually reproducing creatures, cells in multicellular organisms fall into two categories: germ-line cells which become sperm or egg for the next generation; and soma cells that make up the body. The soma cells are the ones that die, and the standard answer to "Why?" is "Why not?" The "disposable soma theory" argues that, in fact, our soma cells die because it's of no value to our DNA to have them keep living forever. It is, in essence, a waste of energy. Throughout most of the history of macroscopic, sexually-reproducing organisms, immortal organisms would not have had an evolutionary advantage. Rather, there was an evolutionary pressure toward organisms that could evolve faster. If a species is going to evolve rapidly, it's valuable for it to have a relatively quick turnover from one generation to the next.

There doesn't seem to be any single cellular "grim reaper" process causing soma cell senescence. Rather, it would appear that there several distinct mechanisms, all acting in parallel and in concert.

There are junk molecules, accumulating inside and outside of cells, simply clogging up the works. Then there are various chemical modifications that impair the functioning of molecular components such as DNA, enzymes, membranes and proteins. Of all these chemical reactions, oxidation has attracted the most attention, and various anti-oxidant substances are on the market as potential aging remedies. Another major chemical culprit is "cross-linking," the occasional formation of unwanted bridges between protein molecules in the DNA—bridges which cannot be broken by the cell repair enzymes, interfering in the production of RNA by DNA. Cross-linkages in protein and DNA can be caused by many chemicals normally present in cells as a result of metabolism, and also by common pollutants such as lead and tobacco smoke.

As time passes, signaling pathways and genetic regulatory networks within cells can be altered for the worse, due to subtle changes in cellular chemistry. The repair mechanisms that would normally correct such errors appear to slow down over time. "Telomeres," the ends of chromosomes, seem to get shorter each time a cell divides, causing normally suppressed genes to become activated and impair cell function. Finally, the brain processes that regulate organism-wide cell behavior decline over time, partly as a result of ongoing cell death in the brain.

The really frustrating thing about all these phenomena is that none of them are terribly different from other processes that naturally occur within cells, and which cells seem to know quite well how to cure and repair. It would seem that cells have just never bothered to learn how to solve these particular problems that arise through aging, because there was never any big evolutionary advantage to doing so. It seems that we may die not because it would be so terribly hard to engineer immortal cells, but because it was not evolutionarily useful to our DNA to allow us to live forever.

Long Life Through Calorie Restriction?

Shifting from the biological mechanisms up to the whole-human-organism level, one fascinating idea that has emerged from anti-aging research is calorie restriction (CR) – put simply, the idea that if you eat about 70% of what you'd ordinarily want, you'll live longer. You need to eat a healthy diet, rich in vitamins and proteins, but low in calories.

Exactly how and why it works at the molecular level isn't yet known – but the reality of the phenomenon is beyond dispute. CR has been tested extensively in various nonhuman mammals. For instance, mice normally don't live over 39 months, but caloric restriction has produced mice with 56 months lifespan. This corresponds proportionally to a 158 year-old human. And these long-lived mice aren't old and crusty—they're oldster/youngsters, keen-minded, strong-bodied and healthy. Studies on monkeys are currently underway, though this naturally will take a while due to monkeys' relatively long lives.

But before you start restricting your calories too heavily, you should know that the evidence suggests it won't be quite as effective in humans as in mice. Roughly speaking, it seems that: the larger the organism, the less the effect of caloric restriction. So it increases the lifespan of nematode worms a lot, mice a fair bit, dogs only a little, and humans—in his hypothesis—by just a few years.

A few things about the biological mechanisms *are* known. CR increases the ability of the body to repair damaged DNA, and it decreases the amount of oxidative (free radical) damage in the body. It increases the levels of repair proteins that respond to stress, it improves glucose-insulin metabolism, and for some reason not fully understood, it delays age-related immunological decline as well. Basically, among other possible factors, many of the well-known mechanisms of senescence set in more slowly if the body has to process less food over its lifetime. The relation of this line of thinking with anti-aging pharmacology has yet to be thoroughly investigated—it may well be there are medications that work most effectively in coordination with a caloric restriction diet.

Among the biological actors underlying CR's impact on aging are the sirtuins, proteins that are addressed by natural substances like resveratrol (found in red wine, and also available now as a nutritional supplement) and also by new drugs under development by Sirtris, a pharma startup recently acquired by GlaxoSmithKline for $730 million. But the sirtuins are certainly not the whole story; our own AI-based analysis of some data related to CR in mice has suggested a host of other biological actors are involved.

The Paleo Lifestyle

A more palatable, and more thoroughly scientifically grounded, alternative to the CR diet is the "Paleolithic lifestyle" advocated by an increasing set of aging researchers. In essence this involves eating the foods that were commonly eaten by humans before the emergence of civilization, and generally trying to emulate the physical patterns of the pre-civilized lifestyle as closely as possible.

Evolutionary biologist Michael Rose (my colleague in some of my recent work on the genomics of longevity) argues that the Paleo lifestyle may be particularly effective for people over the age of 40 or so, because genes frequently have different effects at different life-stages, and their effects at later ages (after the most typical reproductive age) are generally adapted by evolution at a much slower pace. So, our genes' effects on later life-stages are likely more hunter-gatherer-ish than our genes' effects on earlier life-stages – making a Paleo approach even more beneficial for those of us 40 or over.

It's worth noting that Michael is in this age range himself, and he does walk the walk as well as talk the talk – he keeps to the Paleo diet and lifestyle pretty strictly, and he looks and feels great! I myself have been keeping a sort of quasi-Paleo diet and lifestyle, and have noticed some definite improvements in my feeling of physical well-being also. Whereas when I tried the CR diet for a while a few years back, I found myself feeling pleasingly light-headed and positive, but also relatively devoid of drive and energy and ... Well ... Sometimes annoyingly hungry!!!

Aubrey de Grey and the Seven Pillars of Senescence

Lifestyle and diet solutions have a lot to recommend them – especially the fact that they're something we can do right now to help extend our lives, rather than waiting for scientific advances! But there's a limit to what they can do. To really eliminate the plague of involuntary death, new science and technology will be needed – though exactly what kind isn't yet entirely clear.

Aubrey de Grey, as noted above, has put forth some fairly specific proposals in this regard. Of all the senescence researchers in the world, no other has done as much as Aubrey to raise the profile of the aging problem in the public eye. But as well as spreading the good word about the desirability and viability of radical human life extension, he's also proposed some specific scientific ideas. We don't always agree with his proposed solutions to particular sub-problems of aging, but we do find him invariably rational, insightful, and intellectually adventurous.

De Grey's buzzword is SENS, which stands for Strategies for Engineered Negligible Senescence—a very carefully constructed scientific phrasing for what we've loosely been calling here "anti-aging research." The point of the term is that it's not merely slowing down of aging that he's after—it's the reduction of senescence to a negligible level. He's not trying to achieve this goal via voodoo, we're trying to achieve it via engineering—mostly biological engineering, though nanoengineering is also a possibility. De Grey's website gives a very nice overview of his ideas, together with a number of references.

As part of his effort to energize the biology research community about SENS, de Grey launched a contest called the "Methuselah mouse prize"—a prize that yields money to the researcher that produces the longest-lived mouse of species Mus musculus. In fact there are two sub-prizes: one for longevity, and a "rejuvenation" prize, given to the best life-extension therapy that's applicable to an already partially-aged mouse. There is a complicated prize structure, wherein each researcher who produces the longest-lived mouse ever or the best-ever mouse-lifespan rejuvenation therapy receives a bit of money each week until his record is broken.

De Grey's idea is that, during the next decade or so, it should be possible to come pretty close to defeating senescence within mice—if the research community puts enough focus on the area. Then, porting the results from mouse to human shouldn't take all that much longer (biological research is regularly ported from mice to humans, as they are an unusually suitable testbed for human therapies—though obviously far from a perfect match). Of course, some techniques will port more easily than others, and unforeseen difficulties may arise. However, if we manage to extend human lives by 30 or 40 years via partly solving the problem of aging, then we'll have 30 or 40 extra years in which to help biologists solve the other problems.

Theory-wise, de Grey agrees with the perspective we've given above—he doesn't believe there's one grand root cause of senescence, but rather that it's the result of a whole bunch of different things going wrong, mainly because human DNA did not evolve in such a way as to make them not go wrong. On his website he gives the table shown above of the seven causes of senescence, showing for each one the date that the connection between this phenomenon and senescence first become well-known to biologists—and also showing, for each one, the biological mechanism that he believes will be helpful for eliminating that particular cause.

Seven basic causes—are these really all there is? De Grey opines that, "The fact that we have not discovered another major category of even potentially pathogenic damage accumulating with age in two decades, despite so tremendous an improvement in our analytical techniques over that period, strongly suggests that no more are to be found—at least, none that would kill us in a presently normal lifetime." Let's hope he's right, though of course the possibility exists that as we live longer new effects will be discovered—but if we've put enough resources into the anti-aging program, we should be able to combat those as well.

De Grey's particular breakdown into seven causes is slightly arbitrary in some ways, and others would do the breakdown differently; but his attempt to impose a global order on the panoply of aging-related biological disasters is appealing, and reflects a lot of deep thinking and information integration.

One of these "Seven Pillars of Aging" is something that has come up in my own work applying AI to analyze Parkinson's disease, which I'll discuss below: mutant mitochondria. A deeper look at this case is interesting for what it reveals about the strength and potential weaknesses of de Grey's "engineering"-based approach. The term "engineering" in the SENS acronym is not a coincidence—de Grey came to biology from computer science, and he tends to take a different approach from conventional biologists, thinking more in terms of "mechanical" repair solutions. Whether his approach will prove the best or not remains to be seen. We're not biologists enough to have a strong general intuition on this point, but it's often a good bet to say that variety amongst approaches, not a single orthodoxy, will yield the best results. The mainstream molecular biology community seems to think de Grey's proposed solutions to his seven problems reveal a strange taste; but this doesn't mean very much, as the mainstream's scientific taste may well be mortally flawed. Science, like any human endeavor, has its fashions and trends. What is seen as weird science today may be a commonplace field of study in a decade.

Regarding mitochondrial DNA damage, de Grey's current proposal is to fix it in a rather direct way, by replacing the flawed proteins produced by the flawed mitochondrial DNA. This could work because there is already an in-built biological mechanism that carries proteins into mitochondria: the TIM/TOM complex, which carries about 1000 different proteins produced from nuclear DNA into the mitochondria.

What de Grey proposes is to make copies of the 13 protein-coding genes in the mitochondrial genome, with a few simple modifications to make them amenable to the TIM/TOM mechanism, and then insert them into the nuclear chromosomes. In this way they'll get damaged much more slowly, because the nuclear chromosomes are much more protected from mutations than mitochondrial genes.

Sensible enough, no?

On the other hand, I recall a conversation over dinner a few years back, somewhere in northern Virginia, between Aubrey and Rafal Smigrodzki, the biologist who got me involved in Parkinson's research when he was at University of Virginia. Rafal's worry about moving mitochondrial DNA into the nucleus is that its ordinary operations may depend on other things that are happening outside the nucleus, so maybe after it's moved it won't be able to do its thing properly. In other words, maybe Aubrey's "engineering approach" overlooks too much of the complexity of the biological networks involved. So far as I can tell, the jury is still out on this.

Rafal, however, has his own angle on the problem – he left UVA to work for GENCIA Corporation, a Charlottesville firm developing "mitochondrial gene replacement therapy," via a novel technique called protofection. Protofection allows the removal of bad fragments of mitochondrial DNA, and their replacement by good ones. If this could be successfully done in living human brains – and if mitochondrial DNA damage is indeed a major cause of Parkinson's symptoms – then potentially Parkinson's could be cured via gene therapy that simply repairs the flawed regions of mitochondrial DNA.

Whether de Grey's approach or protofection – or something totally different – is the best approach, no one truly knows right now. Unfortunately, neither currently proposed approach is being amply funded at the moment. We can't know for sure unless the research is done, but the research can't be done without funding. Which is why de Grey's publicity and education efforts are so extremely valuable.

Similarly, each of de Grey's other six categories of aging-related damage is amenable to a number of different approaches— we just need to do the experiments and see which one works better.

Radical Life Extension: What's the Real Bottleneck?

Aging is comprehensible but complex – so many different aspects, each addressable by so many human methods. The human mind boggles when confronted with so many complex, interlocking networks. And this, in fact, is a major problem – and one of the reasons why aging hasn't been cured yet. Biologists have gathered a lot of data -- but the human brain ultimately was not evolved for the integrative analysis of a massive number of complexly-interrelated, high-dimensional biological datasets. We desperately try to cast biological data in a form our human brains can understand effectively: We create data visualizations to ease the application of the 30% of our brain that is customized for visual processing; we develop vocabularies and ontologies to better apply the large portion of our brain customized for linguistics. But there is no portion of the brain customized for generating hypotheses by analyzing biological data. At this stage, the weakest link in the biomedical research pipeline is our human brains' lack of ability to holistically understand the mass of data that has been (and is, every day, being) collected, and use this understanding to design new experiments leading to new understanding. It's a somewhat radical assertion, but we contend that the *human brain* is the primary bottleneck along our path to radical life extension.

There are three evident solutions to this problem: improve the human brain, augment it with external tools, or replace it with something better.

The former is an exciting possibility, which will surely be possible at some point, but neuroscience and neuroengineering are currently a long way from enabling robust human cognitive enhancement. Furthermore, advancing neuroengineering is largely a biology problem – which means that a major bottleneck along the path to its achievement is precisely the problem we're talking about, the limitations of the human brain at grappling with masses of biological data.

External tools for biological data analysis are critical and fortunately they are now plentiful, but it is increasingly clear that the sorts of tools we have created are not sufficient to allow us to grapple with the patterns in the data we've collected. Contemporary bioinformatics analysis and visualization software represents a noble, yet ultimately inadequate attempt to work around the shortcomings of the human brain.

To see this, consider the commonplace observation that most geneticists focus their research on a handful of genes, or at very most a handful of biological pathways. This cognitive strategy on the part of researchers makes sense because the human brain can handle only so much information. There are some genes, p53 for example, about which so much information is known that very few human scientists today have it all in their heads. On the other hand, it's also well known that the human body is a highly complex system whose dynamics are dominated by subtle nonlinear interactions between different genes and pathways. So the correct way to analyze biological data is not to focus on individual genes and pathways, but to take more of a holistic, systems-biology approach.

Can software tools help with this? It turns out the answer is yes – but only to a limited extent. While not as commonly utilized as they should be, there do exist statistical and machine learning approaches to analyzing biological data, which take a holistic approach and extract global patterns from huge datasets. Unfortunately, though, these software programs only go so far; they produce results that still need to be interpreted by human biologists, whose expertise is invariably limited in scope, due to the limitations of human memory.

Visualization tools help a lot here as well, but also have fairly strict limitations: the human eye can only take in so much information at one time. It evolved for scanning the African veldt, not the intricacies of biomolecular systems. Even if you had a holographic simulation of some portion of the human body at some particular scale, this still wouldn't allow the human perceptual system to "see the whole," to grasp all the mathematically evident patterns in the data being visualized.

From a scientific perspective, it would be ideal if we could simply replace human biologists with AI systems customized for biological data analysis – systems with the human capability for insight and interpretation (or even more so), but more memory and more capability for quantitative precision, and pattern-analysis capability tuned for biological data rather than recognizing predators on the veldt. Unfortunately, this kind of "AI scientist" does not exist at present. There are serious research programs underway with the aim of producing this kind of software; and an increasing confidence in the AI field that this is indeed an achievable goal[43]. But life extension research, and biological research in general, cannot afford to wait for computer scientists to produce powerful AI – there's too much urgency about moving ahead with solving medical problems causing human suffering, right now. So what we need is a combination of narrow-AI applied to biomedical problems, and development of advanced AGI with a goal toward eventually doing a radically better job.

AI Against Aging: Tales from the Research Front

Having sketched a general picture of the state of longevity biology these days, I'll now get a bit more personal, and discuss some of the specific work I've done applying narrow AI – not AGI yet – to help unravel the biology of aging. As well as uncovering some intriguing aspects of longevity biology, this work has helped me to much more fully understand exactly how AGI would be able to help us figure out how to extend life.

The application of narrow AI technology to biological data in general – and even aging-related data in particular – is a relatively mainstream pursuit now, being carried on by a host of researchers at various universities, companies, and government labs. My own work with my colleagues at Biomind LLC constitutes only a small part of this overall exciting research thrust. But it will give you a concrete flavor for what the field is like these days – just before the dawn of the age of AGI.

---

[43] Goertzel BN, Coelho LS, Pennachin C, Goertzel I, Queiroz M, Prosdocimi F, Lobo F. Learning Comprehensible Classification Rules from Gene Expression Data Using Genetic Programming and Biological Ontologies. In: 7th International FLINS Conference on Applied Artificial Intelligence: 2006; Genova, Italy; 2006.

Overall, my work in narrow AI based biomedical informatics so far has focused on a sort of midway point between the approaches of "better tools" and "replace the humans". As you already know if you've been paying any attention to the earlier parts of this book, I believe that the creation of a powerful AI scientist is a real possibility within the next few decades, maybe even within the next decade – and I'm spending a lot of my time working specifically toward this goal. But, via my role in the bioinformatics firm Biomind LLC, which is working with the NIH and has worked with the CDC and various academic biomedical labs, I'm also acutely aware that there is important biomedical data being generated right now about important human problems, and we've got to deal with it as best we can. So the approach my colleagues and I are taking in the bio space is an incremental one: as our ambitious AI scientist is gradually created (and it's a long-term research project, as one might expect), we are utilizing the various modules of the overall AI system to analyze biological datasets. Of course, the AI modules are not as powerful as a full-scale AI scientist would be, but our experience has shown that they can still provide insights beyond what human scientists can achieve unaided, or using conventional tools. In this way AI and biomedical science can progress together: The more progress we make toward the AI scientist, the more powerful the insights generated by the partial versions of the system.

AI Uncovers the Role of Mitochondrial DNA in Parkinson's and Alzheimer's Disease

One of the most exciting chapters so far in my exploration of the application of AI to bioscience, involved work we did in 2005 analyzing data regarding the genetic roots of Parkinson's disease. In this case, the result of the AI analysis was a powerful statistical validation of the hypothesis that Parkinson's is caused by mitochondrial mutations. These results seem reasonably likely to lead to a practical diagnostic test for Parkinson's, and if the work being done at Gencia[44] on protofection works out, they may ultimately form the foundation of a mitochondrial gene therapy based cure.

Over a million Americans have Parkinson's disease. Yet in spite of years of effort by medical researchers, tracking down the genetic roots of the disorder has proved devilishly difficult. The DNA one usually hears about lies in the nucleus of a cell, the cell's center. In many cases the genetic roots of disease can be traced down to mutations in the nuclear DNA, called SNP's or Single-Nucleotide Polymorphisms. Biomind had a significant success with this sort of analysis when analyzing SNP data regarding Chronic Fatigue Syndrome: the AI was able to tease out patterns of mutational combination that provided the first real evidence that CFS is at least partially a genetically-founded disease[45]. While this sort of approach has not proved workable for Parkinson's, a variation proved dramatically successful. Mitochondria, the cell's energy-producing engines, also contain a small amount of DNA. What the AI has told us is that the right place to look for the genetic roots of Parkinson's is in the mutations in the *mitochondrial* DNA. Our software identified a particular region of a particular gene on the mitochondrial genome that appears to be strongly associated with Parkinson's disease[46].

Much smaller, lesser known and lesser studied than its nuclear cousin, the mitochondrial genome is nonetheless vital to cellular

---

[44] Smigrodzki RM, Khan SM. Mitochondrial microheteroplasmy and a theory of aging and age-related disease. Rejuvenation Res. 2005 Fall;8(3):172-98.

[45] Goertzel BN, Pennachin C, de Souza Coelho L, Maloney EM, Jones JF, Gurbaxani B. Allostatic load is associated with symptoms in chronic fatigue syndrome patients. Pharmacogenomics 2006; 7:485-494.

[46] Smigrodzki R, Goertzel B, Pennachin C, Coelho L, Prosdocimi F, Parker WD, Jr. Genetic algorithm for analysis of mutations in Parkinson's disease. Artificial Intelligence in Medicine 2005; 35:227-241.

function in humans and other animals. The human mitochondrial genome only contains seven genes, whereas the nuclear genome contains around 30,000 at last count. But these seven genes carry out a lot of valuable functions. If they stop working properly, serious problems can arise. In 1999, Dr. Davis Parker, together with Russell H. Swerdlow and scientists from San Diego firm MitoKor's published work suggesting that defects in the mitochondrial genome may be correlated with Parkinson's disease. As a baby's mitochondrial DNA comes entirely from its mother, these results suggest that Parkinson's may be passed maternally – but that its defects can skip generations, making the emergence of the disease appear random.

The work Parker and Swerdlow's team did involved clever manipulations of embryonic human nerve cells. They removed the mitochondrial DNA from the embryonic nerve cells and replaced it with other DNA: Sometimes from healthy people and sometimes from Parkinson's patients. What resulted was the nerve cells receiving the mitochondrial DNA from Parkinson's patients began acting like nerve cells on MPTP. Low complex I activity, meaning insufficient energy obtained from mitochondria – and eventually leading to Parkinson's-like sluggishness.

These results were fascinating and suggestive – but where were the actual mutations? All this showed was that the problem lay somewhere in the mitochondrial genome. The question was where. Which mutations caused the problem?

To answer this question, Parker and colleagues sequenced mitochondrial DNA drawn from the nerve cells of a number of Parkinson's patients, as well as a number of normal individuals, and looked for patterns. But to their surprise, when in 2003 they set about seriously analyzing this data, they found no simple, consistent pattern. There were no specific genetic mutations common to the Parkinson's patients that were not common to samples taken from healthy subjects.

Enter artificial intelligence. Dr. Rafal Smigrodzki, one of Parker's collaborators, was familiar with my AI research work and suggested that perhaps my AI technology might be able to find the patterns in the mitochondrial DNA data.

To make a long story short, it worked. Appropriately enough, the solution turned out to be an AI software technique called "genetic algorithms," which simulates the process of evolution by natural selection – beginning with a population of random solutions to a problem, then gradually "evolving" better solutions via letting the "fittest" solutions combine with each other to form new ones, and making small "mutations" to the fittest solutions. In this case, what the software was "evolving" was potential patterns distinguishing Parkinson's patients from healthy subjects based on the sequences of amino acids in their mitochondrial DNA. This kind of data analysis is highly exploratory and is never guaranteed to yield a solution – but in this case things worked out happily, and a variety of different data patterns were discovered.

The trick, it turns out, is that while there are no specific mutations corresponding to Parkinson's disease, there are regions – and combinations of regions -- of the mitochondrial genome that tend to be mutated in Parkinson's patients. There are many different rules of the form "If there are mutations in this region of this mitochondrial gene and that region of that mitochondrial gene, then the person probably has Parkinson's disease." While it took some advanced AI technology to find these patterns, once discovered, the patterns are very easy for humans to understand. The patterns were validated by subsequent biological analysis on additional patients[47].

---

[47] Smigrodzki R, Goertzel B, Pennachin C, Coelho L, Prosdocimi F, Parker WD, Jr. Genetic algorithm for analysis of mutations in Parkinson's disease. Artificial Intelligence in Medicine 2005; 35:227-241.

Yet more excitingly, we've done further work (to be published shortly) with Dr. Parker on comparable data regarding Alzheimer's disease, showing patterns that are similar in nature but different in detail. Once again, although the crucial idea to look at the mitochondria in the first place was provided by human biological intuition, the human brain was unable to detect the relevant patterns in the mitochondrial mutation data, even augmented with cutting-edge statistical tools. But AI found the relevant patterns, which are then easily validated via further biological experiments.

AI Helps Unravel the Genetic Mechanisms Underlying the Efficacy of Calorie Restriction for Life Extension

As well as helping to understand and diagnose (and, ultimately, cure) aging-related diseases like Parkinson's and Alzheimer's, AI technology can help us better understand, refine and design methods for extending the maximum lifespan of organisms. One recent example of this is a study my colleagues and I recently published in Rejuvenation Research[48], pertaining to the genetic mechanisms underlying the impact of calorie restriction diets on maximum lifespan. The exact mechanism by which calorie restriction works remains incompletely understood (though there are plenty of theories!), but our AI-based analysis revealed a central role for several genes whose involvement in CR's efficacy was not previously known. These results suggest a number of specific biological experiments, and we are in discussions with biology research labs regarding the best way to carry out these experiments. These experiments of course will produce new data to be analyzed via AI algorithms, and will likely provide information on how various elements of the many existing theories of CR's efficacy combine to provide the true explanation. Through this sort of iterative interaction between AI

[48] Goertzel B, Coelho L, Mudado M. "Identifying the Genes and Genetic Interrelationships Underlying the Impact of Calorie Restriction on Maximum Lifespan: An Artificial Intelligence Based Approach", to appear in Rejuvenation Research.

analysis, human judgment and laboratory experiments, we can progress much faster than would be possible without AI in the picture.

In our application of AI to CR, we initially fed our AI system three datasets that other researchers had posted online, based on their work studying mice on calorie restriction diets. We then merged these three datasets into a single composite dataset for the purpose of conducting a broader-based analysis, using AI technology rather than the standard statistical methods that the researchers had originally used on their datasets.

Along with providing a large amount of other information, this analysis resulted in a list of genes that the AI found to be important for CR's impact on lifespan. An essential point here is that the AI was capable of teasing out nonlinear interactions between different genes and gene products. The genes that the AI points out as important for CR and its impact on aging are important, not necessarily in terms of their individual actions, but most often largely in terms of their interactions with other genes.

The AI also provided a map of gene interrelationships (shown in Figure 1), suggesting which *inter-gene interactions* are most important for the effect of CR on life extension. In particular, our graphical analysis revealed that the genes Mrpl12, Uqcrh and Snip1 play central roles in the effects of CR on life extension, interacting with many other genes (which the analysis enumerates) in carrying out their roles. This is the first time that the genes Snip1 and Mrpl12 have been identified as important in the aging context.

To double-check the validity of these results we obtained from analyzing three datasets at once, we then ran the same AI processes all over again, but throwing a fourth dataset into the mix. Much to our relief the results were largely the same – suggesting that the AI is producing real biological insights, not just some kind of data processing artifacts.

Broadly, the biological interpretation of these analytical results suggests that the effects of CR on life extension are due to multiple factors, including factors identified in prior theories of aging, such as the hormesis[49], development[50], cellular[51] and free radical[52] theories. None of these individual theories stands out as obviously correct, based on the patterns of gene-combination effects identified by the AI system. But genes with predicted involvement according to many of these theories play a role, along with other genes not highlighted by any prior theories or experiments.

---

[49] Sinclair DA. Toward a unified theory of caloric restriction and longevity regulation. Mech Ageing Dev 2005; 126:987-1002.

[50] de Magalhaes JP, Church GM. Genomes optimize reproduction: aging as a consequence of the developmental program. Physiology (Bethesda) 2005; 20:252-259.

[51] Shay JW, Wright WE. Hallmarks of telomeres in ageing research. J Pathol 2007; 211:114-123.

[52] Harman D. Aging: a theory based on free radical and radiation chemistry. J Gerontol 1956; 11:298-300.

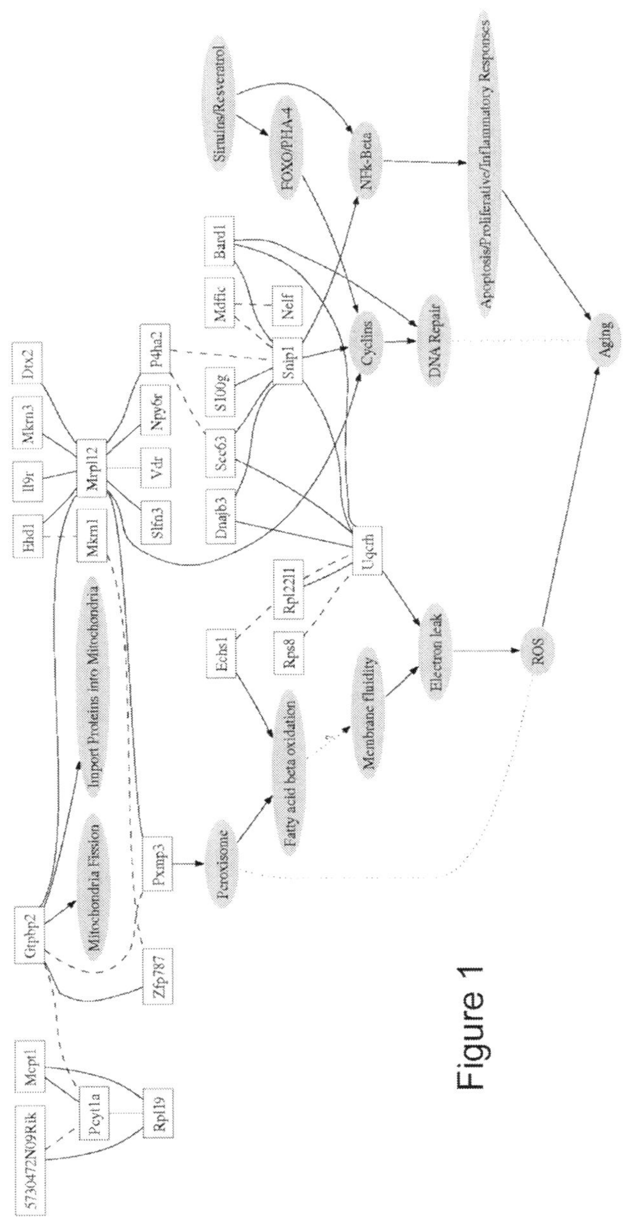

5730472N09Rik
Mcpt1
Pcyt1a
Rpl19
Gtpbp2
Zfp787
Mitochondria Fission
Import Proteins into Mitochondria
Pxmp3
Ehd1
Mkrn1
Il9r
Mrpl12
Mkrn3
Dtx2
Slfn3
Vdr
Npy6r
P4ha2
Peroxisome
Fatty acid beta oxidation
Echs1
Rps8
Rpl22l1
Membrane fluidity
Electron leak
ROS
Dnajb3
Sec63
S100g
Snip1
Mdfic
Nelf
Bard1
Uqcrh
Cyclins
DNA Repair
Aging
Sirtuins/Resveratrol
FOXO/PHA-4
NFk-Beta
Apoptosis/Proliferative/Inflammatory Responses

Figure 1

Unraveling the Mystery of the Methuselah Flies

Perhaps my most exciting current research project involving AI and biology data has to do with Michael Rose's Methuselah flies, mentioned above: Fruit flies that have been bred by directed evolution, over the last 30 years, to live 5x longer than ordinary fruit flies. Simply by setting up a situation where longer-lived flies are more likely to breed with each other, and letting it operate for many, many generations, a new strain of flies was created. This is a miracle, and a puzzle: because these Methuselah flies were created via directed evolution rather than genetic engineering or other "direct" methods, we don't know how what it is that makes them live so long. Now we need to "reverse engineer" what directed evolution did, and understand what combination of genetic mutations occurred to create the long-lived flies, and why these mutations had the impact they did. This is not a completely simple matter, because evolution is messy: the Methuselah flies are bound to have a lot of inessential differences from regular flies along with the functionally critical ones, and the inessential and critical ones are going to be complexly bound up with each other. Traditional statistical analysis methods can identify some genes that are important to understanding the difference between the Methuselah flies and ordinary flies, but, they can't unravel the genomic, proteomic and metabolomic interrelationships.

Even without a full understanding of what keeps them ticking, analysis of the Methuselah flies has borne some fruit (sorry!). Genescient, the company that now has rights to the IP of the Methuselah flies, has used the Methuselah flies to find some substances that can be fed to normal flies to make them live much longer than usual. Furthermore, this research has led to insights regarding nutraceuticals for promoting longevity in humans. But these results are minor compared to what could be achieved if the essential cause of the Methuselah flies' longevity were understood. Not all biologists agree that understanding aging in fruit flies will help us understand human aging, but there's a strong argument to be made. To the extent that aging is a basic property of cellular function, it is likely to be the same process across many organisms – and indeed, Genescient has done studies showing that the genes most significant in characterizing the Methuselah flies tend to be ones that also relate to human diseases.

Aubrey de Grey's "engineering approach" to combating aging focuses on the symptoms of aging, which occur at various levels throughout the body (as a single example, he proposes to use certain bacteria to clean up "gunk" that appears between the body's cells increasingly with age). While this approach may have great value, there's also something to be said for trying to fix the basic cellular processes underlying aging. Perhaps if these processes are fixed then many of the symptoms will disappear on their own. Ultimately, de Grey's approach and Genescient's approach may lead to complementary therapies.

So far we have only applied AI to a small set of fly data, but we have already found some interesting conclusions. The general role of AI here is to identify which genes are important for the Methuselah flies' longevity, and how these genes combine with each other – and based on this understanding, to figure out which pathways can be impacted with pharmaceuticals or nutraceuticals to cause ordinary flies to live longer. AI can also select among the various relevant genes and pathways to estimate which ones are most likely to lead to human aging therapies. As in our previous examples, the AI is far from autonomous here; it is serving as a helper to human biologists and data analysis. But there is a lot of data and the biology is complex, so the latter can use all the help they can get!

I can't recount the details results of my work with Genesicent here due to intellectual property concerns, but I can review the basic sorts of things we're finding. For instance, we have found one gene that seems to be very important to fly longevity, and that produces a certain enzyme known in humans due to its deficiency in people with a certain monogenetic disease involving central nervous system malfunction. Another gene emerging as important is a tumor suppressor gene (and the relation between cancer tumor suppression and aging is very well known), which plays a role in the Methuselah flies in combination with several particular genes related to metabolism. None of these findings, in itself, tells you why the Methuselah flies live so long – but they point research in specific directions, some of which would not have been conceived based on this data without the AI-based analysis results.

AI That Reads Biological Research Articles

So far we've been discussing the use of AI to analyze quantitative biological datasets. But there's another fact that must also be considered, which is that the vast majority of biomedical knowledge online right now exists only in textual format. Most datasets aren't placed online, and as big as the biological databases are, most knowledge that could be placed in there, actually hasn't been, either because no one has gotten around to it, or because researchers prefer to keep their data proprietary.

For example, at Biomind we've done a lot of work with the Gene Ontology, which is an outstanding database that categorizes genes by function. If you look up "apoptosis" in the Gene Ontology, you'll find a few dozen genes that have been categorized as being associated with apoptosis – preprogrammed cell death. But the catch is, if you browse through the journal literature online, you'll find even more. The Gene Ontology can't keep up. This is a tribute to the rapid pace of biomedical research these days, but it's also an indication of one direction biomedical software has got to go in: We've got to write computer programs that can grab the information directly from the texts where it's been published! This is a domain of research called Bio-NLP – bio natural language processing.

Once a sufficiently powerful AI scientist is created, Bio-NLP won't be necessary, as the AI will simply recognize all the relevant patterns in the data directly, without need for human insight. But we're not there yet. So at the present time, the best strategy for AI data analysis is to incorporate all available sources of information, including direct experimental data and text humans have produced based on interpreting that data.

In 2006, I co-organized the sixth annual Bio-NLP workshop, as part of the annual HTL-NAACL Computational Linguistics conference. At previous Bio-NLP workshops, nearly all the work presented had pertained to fairly simple problems, such as recognizing gene and protein names in research papers (a task made more difficult than it should be by the presence of multiple naming conventions among biologists). But starting in 2006 we saw more and more researchers creating software with the capability to recognize *relationships* between biological entities, as expressed in natural language text; and this trend has intensified subsequently. The latest Bio-NLP software (see Rzhetsky's work [53] for an impressive example) takes in a research paper and tells you which genes, proteins, chemical and pathways are mentioned, and how they are proposed by the authors to relate to each other (which genes are in which pathways, which enzymes catalyze which reactions, which genes upregulate which others, etc.). This is a far cry from full understanding of the contents of research papers, but it's definitely a start.

AI Based Logical Inference Based on Information Automatically Extracted from PubMed Abstracts

The paper I presented at Bio-NLP 2006 regarded a research prototype called BioLiterate, which we built for the NIH Clinical Center in 2005. What the BioLiterate prototype did was extract relationships from various biomedical research abstracts, and try to glue them together using logical reasoning. So, for example, if one paper said that p38 map kinase inhibition prevents bone

---

[53] Krauthammer M, Rzhetsky A. GENIES: a natural-language processing system for the extraction of molecular pathways from journal articles. Bioinformatics 2001;17(Suppl 1):S74–82 (0)

loss, and another paper said the DLC inhibits p38, then the software would put A and B together, deciding (using logical reasoning) that maybe DLC prevents bone loss (the actual sentences the AI used in these inferences, found in PubMed abstracts, are shown in the figure above). The logical inference was provided by the Probabilistic Logic Networks module of the Novamente Cognition Engine[54]. BioLiterate was a prototype, rather than a robust and deployable software solution, but it made its point: If you build a Bio-NLP system and then use the right sort of rules to pipe its output into a computational reasoning system, you get an automated biological hypothesis making system.

The Holistic Biobase

The work we've discussed above, applying AI to bioinformatics, has already led to exciting results. A growing community of other researchers, doing similar work, is finding similar successes. Continuing this general approach, applying AI technology to various datasets in isolation or in small groups, there is little doubt that an ongoing stream of comparable results can be obtained, providing a significant and worthwhile acceleration to the advancement of bioscience.

But, we could do a lot better. The real future of bioscience, we suggest, lies in the simultaneous analysis of a lot more than the four datasets we considered in our calorie restriction study. We need to feed dozens, hundreds, thousands, tens and hundreds of thousands of datasets simultaneously into the same AI system – along with all the biological texts online – and let the AI go to town hunting down the patterns that are concealed therein. AI can detect far more patterns in such a data-store than the human mind.

---

[54] Ikle M, Goertzel B, Goertzel I, Heljakka A. "Probabilistic Logic Networks: A Comprehensive Framework for Uncertain Inference", Springer, May 2008.

Right now, the mass of available data is terrifyingly underutilized, due to the limitations of the human brain and the corresponding processes of the scientific community (which are adapted to the limitations of the human brain). Human scientists analyze individual datasets, or small collections of datasets, using brains that evolved for solving other sorts of problems (aided by statistical, visualization and in rare cases AI tools); and then these humans write papers summarizing their results. Of course, the papers written about a certain dataset ignore nearly all the information in that dataset, focusing on the particular patterns that the researchers noticed (which are often the ones they were looking for in the first place, based on their prior knowledge and biases). Then, researchers read the papers other researchers have written, and use the conclusions in these papers to guide the analysis of new datasets. The multiple datasets that have been collected are brought together indirectly only via human beings reading and writing papers, each of which contains an extremely partial view into the data on which it's based. This is a dramatic, tragic loss of information compared to what would happen if the datasets were actually analyzed collectively in a serious way.

What we're suggesting is the creation of a ***Holistic Biobase*** – a massive data repository containing all the biomedical information on the Web today – including quantitative data, relational data, textual information in articles and abstracts... Everything. The data in this repository should then be analyzed using powerful AI systems that are able to study the data as a whole, identifying complex patterns not amenable to direct human analysis nor conventional statistics. These software systems will help humans make better discoveries, and in some cases they will surely make new discoveries on their own – suggest new experiments, propose new hypotheses, make connections that no human could make due to our limited ability to store and analyze information in our brains.

The Holistic Biobase should ideally be an open information resource, so that any scientist with statistical or AI tools and a bit of savvy can crunch the data in their own way. A decent, if partial, model for the Holistic Biobase is Freebase[55], which is an open online database containing various sorts of information of general interest. In principle, one could just load biological datasets into Freebase, but in practice this isn't likely to be the best approach, for several reasons. Freebase is a traditional relational database, which is not the most natural data structure for AI purposes (a graph database would be preferable). And more critically, it doesn't solve the problems of metadata standardization and data normalization, which are perhaps the main obstacles standing in the way of constructing the variety of mega-bio-database I'm envisioning.

If the Holistic Biobase concept sounds overambitious and fanciful, please remember that the Human Genome Project once sounded very much the same. A few decades ago the "synthetic organism" project of Venter's lab at the J. Craig Venter Institute in Rockville, Maryland would also have sounded science-fictionally speculative. And how many people would have labeled the notion of a Google-scale database of online documents implausible or insane, just one or two decades ago? Biology and computer science both are in the midst of phases of rapid advance, which opens up possibilities that could barely have been conceived of before.

As a very simple example of the value the Holistic Biobase would have, let's turn back to the calorie restriction data analysis project mentioned above. We're excited with the results we achieved based on our four-dataset analysis – but it's easy to see how much more powerful the results could be if we had a massive integrative data repository at our disposal. For example, calorie restriction is connected with energy metabolism, a connection we as humans can exploit by interpreting the results of calorie restriction data in the context of our own knowledge

---

[55] Freebase: an open, shared database of the world's knowledge. http://www.freebase.com

about energy metabolism pathways. But what if we integrated masses of raw data regarding energy metabolism in various aging-related contexts into the analysis – and looked at this data together with the calorie restriction data? Who knows what might turn up? Bodies are complex systems, and the effect of calorie restriction on life extension is surely not a phenomenon best understood in isolation. And of course there are a dozen other pathways that should be considered along with energy metabolism.

What kinds of AI algorithms will be able to grapple with the Holistic Biobase in a really effective way? We don't have much experience doing this kind of massive-scale biological data analysis, but the experience we do have gives us significant guidance. There have already been some commercial products pushing in this direction – for instance Silicon Genetics' GeNet database (for microarray data) and associated MetaMine statistical datamining package. But GeNet/Metamine handles only standard statistical methods, and applies only to microarray data. On the other hand, the methods we've been using in Biomind to date are more advanced analytically and are oriented toward combined analysis of multiple types of data. However, they have not yet been tailored for massive-scale data analysis.

Our strong suspicion is that to handle the Holistic Biobase, new methods will be needed. Current applications of AI to bioinformatics have focused on the application of machine learning algorithms for pattern recognition – essentially, algorithms that look at one or more datasets and explicitly scan them for patterns using complex algorithms. To handle larger numbers of data and yet preserve the capability for analytical sophistication, a paradigm shift will be required – and this paradigm shift ties in naturally with the trends of development in the AI field itself. What is needed is the fusion of bioinformatics data analysis with *automated reasoning*. More specifically: Automated *probabilistic* reasoning, since biological data is riddled with uncertainties. Automated reasoning allows an AI system to study a handful of datasets, derive results regarding the patterns in these datasets, and then extrapolate these patterns to see what they imply about other datasets. This step of inferential extrapolation allows far more scalable analysis than machine-learning pattern-recognition methods alone. My own team is currently pursuing this vision via integrating our OpenBiomind bioinformatic AI software with our OpenCog general-AI platform, which includes a powerful probabilistic inference framework called Probabilistic Logic Networks[56]. This will allow us, for example, to massively extend our calorie restriction data analysis project, to include numerous datasets drawn from studies of different but allied biological phenomena.

While this vision goes a fair bit beyond current practice, there are some contemporary projects with smaller but vaguely similar ambitions. One example is a project called ImmPort – this is a program funded by the National Institute of Health, specifically the National Institute for Allergies and Infectious Diseases, which Biomind is involved with via a subcontract to Northrop-Grumman IT. ImmPort is still in the making, but what it's going to be, when it's finished, is a Web portal site for NIH-funded immunologists. Biomind has played a small role in this – to integrate bioinformatics analysis technologies into the portal, both innovative machine learning techniques and more standard methods. The most exciting part of ImmPort is probably its potential to enable massive data integration. When an immunologist uploads data into ImmPort, it will automatically be put in a standard format, so it can be automatically analyzed in the same way as all the other datasets that were uploaded – and, most excitingly, so it can be analyzed in terms of the patterns that emerge when you put it together with all the other datasets. This is something that's hardly ever being done right now – the application of bioinformatic technology to look for patterns spanning dozens or hundreds or thousands of datasets.

---

[56] Ikle M, Goertzel B, Goertzel I, Heljakka A. "Probabilistic Logic Networks: A Comprehensive Framework for Uncertain Inference", Springer, May 2008.

However, the scope of AI currently envisioned within ImmPort is restricted to machine-learning algorithms; extension to more powerful automated inference methods is beyond the scope of the project.

Projects like ImmPort are definitely a step in the right direction – but only a step. Even if every immunologist on the planet were to upload their data into ImmPort, and even if ImmPort were to incorporate inference-based data analysis, the restriction to immunological data alone would still constitute a huge limitation. The immune system is not an island, it is intricately connected with nearly all other body systems. As an example, in its work with the CDC, Biomind found that Chronic Fatigue Syndrome is most likely a complex interaction between immune, endocrine, autonomic nervous and other functions. What we need is not just a holistic immunology database but a holistic biology database, and with a focus on powerful cross-dataset AI analysis as well as statistical and machine learning methods. Furthermore, there is as yet no ImmPort analogue for data directly related to life extension.

Our hope is that over the next decade the ideas described here will become boring and mainstream, and the value of massive, sophisticated, AI-based cross-dataset analysis will move from outrageous to obvious in the consensus view. Until that time, we will continue to be slowed down in our quest to extend human life and cure human disease by the limitations of our human brains at analyzing the relevant biological data.

***And a post-script from 2014**: The ideas in the above essay have NOT yet become boring and mainstream. But they are not as outrageous-sounding as they were in 2008, that's for sure. And we are gradually making progress toward a Holistic Biobase, via integrating various biological knowledge bases and datasets into the OpenCog Atomspace. The vision articulated in "AI Against Aging" seems to be slowly moving toward realization*

– though, as is often the case with technology I'm interested in, much more slowly than one would hope, due to funding and attention issues.

17

AGI and Magical Thinking

In March 2010, while the nuclear reactor disaster in Fukushima Japan was all over the news, a friend asked me to write him a few paragraphs on the topic of AGI and nuclear disasters. How, he asked me, might disasters like Fukushima be prevented if we already had powerful artificial general intelligence?

I actually didn't think this was a great idea for an article, because it seemed too obvious to me that problems like Fukushima could easily be avoided if one had radically superhuman intelligence at one's disposal (although, new and currently unforeseen problems might well arise!). But he insisted, and said he had a particular group of colleagues he wanted to distribute my thoughts to... So I went ahead and wrote him the paragraphs. And having done that, I figured I might as well post the result as a brief item in H+ Magazine.

The H+ post received a fair bit of negative attention, even to the point of getting slammed by Forbes Magazine blogger Alex Knapp (whom I emailed with a bit afterwards, and who turned out to be a really nice guy with a good knowledge of science). Overall the discussion with the article's critics was interesting and highlighted the difficulties of getting people – even highly intelligent and educated people – to really understand the power that advanced AGI would have.

First, the little article in question...

Could AGI Prevent Future Nuclear Disasters?

Ben Goertzel; March 23, 2011

In the wake of a tragedy like the nuclear incidents we're currently seeing in Japan, one of the questions that rises to the fore is: What can we do to prevent similar problems in the future?

This question can be addressed narrowly, via analyzing specifics of nuclear reactor design, or by simply resolving to avoid nuclear power (a course that some Western nations may take, but is unlikely to be taken by China or India, for example). But the question can also be addressed more broadly: What can we do to prevent unforeseen disasters arising as the result of malfunctioning technology, or unforeseen interactions between technology and the natural or human worlds?

It's easy to advocate being more careful, but careful attention comes with costs in both time and money, which means that in the real world care is necessarily compromised to avoid excessive conflict with other practically important requirements. For instance, the Japanese reactor designs could have been carefully evaluated in scenarios similar to the one that has recently occurred; but this was not done, most likely because it was judged too unlikely a situation to be worth spending scarce resources on.

What is really needed, to prevent being taken unawares by "freak situations" like what we're seeing in Japan, is a radically lower-cost way of evaluating the likely behaviors of our technological constructs in various situations, including those judged plausible but unlikely (like a magnitude 9 earthquake). Due to the specialized nature of technological constructs like nuclear reactors, however, this is a difficult requirement to fulfill using human labor alone. It would appear that the development of advanced artificial intelligence, including Artificial General Intelligence (AGI) technology, has significant potential to improve the situation.

An AI-powered “artificial nuclear scientist” would have been able to take the time to simulate the behavior of Japanese nuclear reactors in the case of large earthquakes, tidal waves, etc. Such simulations would have very likely led to improved reactor designs, avoiding this recent calamity plus many other possible ones that we haven’t seen yet (but may see in the future).

Of course, AGI may also be useful for palliating the results of disasters that do occur. For instance, cleanup around a nuclear accident area is often slowed down due to the risk of exposing human workers to radiation. But robots can already be designed to function in the presence of high radiation; what’s currently underdeveloped is the AI to control them. And, most of the negative health consequences of radiation from a nuclear accident such as the recent ones are long-term rather than immediate. Sufficiently irradiated individuals will have increased cancer risk, for example. However, the creation of novel therapies based on AI modeling of biological systems and genomic data, could very plausibly lead to therapies remedying this damage. The reason relatively low levels of radiation can give us cancer is because we don’t understand the body well enough to instruct it how to repair relatively minor levels of radiation-incurred bodily damage. AGI systems integratively analyzing biomedical data could change this situation in relatively short order, once developed.

Finally, the creation of advanced intelligences with different capabilities than the human mind, could quite plausibly lead to new insights, such as the development of alternative

> power sources without the same safety risks. Safe nuclear fusion is one possibility, but there are many others; to take just one relatively pedestrian example, perhaps intelligent robots capable of operating easily in space would perfect some of the existing designs for collecting solar energy from huge solar sails.
>
> There is no magic bullet for remedying or preventing all disasters, but part of the current situation seems to be that the human race's ability to create complex technological systems has outstripped its ability to simulate their behavior, and foresee and remedy the consequences of this behavior. As the progress of technology appears effectively unstoppable, the most promising path forward may be to progressively (and, hopefully, rapidly) augment the human mind with stronger and stronger AI.

OK -- now, I admit this wasn't my strongest article ever. It was hastily typed in a few minutes, more of a quickie blog post than a real magazine article. But to me, its weakest point was its *obviousness*. Alex Knapp's Forbes critique, however, took a quite different tack, accusing me of unscientific "magical thinking." In his own words:

> *... You can show the reliance on magical thinking with just a few quick word changes. For example, I'm going to change the title of the article to "Could* ***Djinn*** *Prevent Future Nuclear Disasters?", then make just a handful of word changes to the paragraphs quoted:*
>
> *What is really needed, to prevent being taken unawares by "freak situations" like what we're seeing in Japan, is a radically lower-cost way of evaluating the likely behaviors of our technological constructs in various situations, including those judged plausible but unlikely (like a magnitude 9 earthquake). Due to the specialized nature of technological constructs like nuclear reactors, however, this is a difficult requirement to fulfill using human labor alone. It would*

> *appear that* ***finding magic lamps that hold Djinn*** *has significant potential to improve the situation.*
>
> ***A Djinn*** *would have been able to take the time to simulate the behavior of Japanese nuclear reactors in the case of large earthquakes, tidal waves, etc. Such simulations would have very likely led to improved reactor designs, avoiding this recent calamity plus many other possible ones that we haven't seen yet (but may see in the future).*

Very funny, right? Three cheers for Alex's sleight of word!

But, as I pointed out in my reply to Alex: What if someone in 1950 had forecast the Internet and everything it can do in 2011?

They could have written "The Internet will help you navigate your car; the Internet will let you select millions of books to read from any house or mobile phone; the Internet will let you reserve flights and monitor global weather conditions; the Internet will help spark revolutions in Third World countries; the Internet will allow people to get college educated from their homes."

And some clever wordsmith could have spoofed them by replacing "Internet" with "Djinn", and accusing them of "magical thinking."

"The Djinn will help you navigate your car; the Djinn will let you select millions of books to read from any house or mobile phone; the Djinn t will let you reserve flights and monitor global weather conditions; the Djinn will help spark revolutions in Third World countries; the Djinn will allow people to get college educated from their homes..."

And most people in 1950 would have laughed along knowingly, at the fools who thought some mythical hypothetical construct called the "Internet" could do all those things.

Of course the Internet has limitations, which would have been hard to foresee in 1950. But nevertheless, it does a lot that would have seemed like magic in 1950.

As Arthur C. Clarke said, "Any sufficiently advanced technology is indistinguishable from magic."

My thinking about AGI may be wrong but it's not "magical"!

Now, I suppose Alex's main point (very loosely paraphrasing) was that just listing amazing things AGI can do isn't very useful, and it would be better to describe a path to actually making these things work. And I agree with that. A detailed discussion of how to get from current tech to robots capable of solving or preventing nuclear disasters would have been a better article!

I never wrote that better article (though I would if I had some evidence the folks in charge of preventing or mitigating nuclear disasters had some propensity to listen to me), but Alex's critique did prompt me to do a little reading on the current state of robotics for nuclear disaster management, which I posted in a P.S. to my original brief article.

I found a very nice article in the IEEE spectrum, describing in detail some of the reasons why current robot technology is only of limited use for helping with nuclear disasters. The basic reasons come down to

- lack of radiation shielding (a plain old engineering problem rather than directly an AGI problem — though hypothetically an AGI scientist could solve the problem, I'm sure human scientists can do so also).
- relative physical ineptitude at basic tasks like climbing stairs and opening doors (problems to be solved by a combination of engineering advances and intelligent control (ultimately AGI) advances)

- the need for tele-operation which is awkward when robots are moving around inside shielded buildings and so forth. This is purely an AGI problem — the whole goal of applying AGI to robotics is to let them operate autonomously

There are many approaches in the research community aimed at creating AGI robots of this sort, and if things go well one of these may lead to robots capable of providing serious aid to nuclear disasters within a, say, 5-25 year timeframe. As well as my own OpenCog approach, one can point to a load of others – say, Juyang Weng's developmental robotics project[57] at Michigan State University, the European IM-CLEVER [58] ("Intrinsically Motivated Cumulative Learning Versatile Robots") project which aims to make a robot capable of autonomously learning various practical tasks, initially childlike ones and then growing more sophisticated. These exemplify the kind of work that I think will eventually lead to robots capable of fixing nuclear disasters – which of course is just the most obvious of the many ways that AGI may help to prevent or mitigate nuclear disasters.

What looks at first glance like magical thinking, turns out in fact to be rational extrapolation from the current work of serious scientists at major universities. Exactly how long it will take to get from here to there is impossible to know rigorously, and the experts disagree – but there's a big difference between "we'll almost surely get there with science and engineering, though we don't know for sure when" and "magical thinking." Nuclear disaster recovery robots are not like femtotech – something that's probably feasible but will require radical new science, and might end up being impossible according to portions of physical law that are now poorly understood. They're more like Drexler-style nanotech – clearly feasible via physical law, and reasonably seen as an extrapolation of current research, though not close

---

[57] http://www.cse.msu.edu/~weng/

[58] http://www.im-clever.eu/

enough to current knowledge as to succumb to straightforward schedule estimation.

Another comment on my brief nuclear disaster piece was posted on an email list by AI researcher Steve Richfield, who argued that:

> *The disaster is the child of two incredible simple human errors:*
>
> 1. *A simple toilet-filling mechanism connected to an outside inlet for use in disasters would have provided a way to keep the nuclear fuel submerged despite technical failures. No one would ever install such a simple thing because it would be an admission of engineering weakness.*
> 2. *Spent fuel is kept on site rather than being reprocessed for purely political reasons – no one wants the reprocessing done in their own back yard.*
>
> *An AGI would be faced with these same two pressures. Only through the exercise of very un-democratic power could it get past them.*
>
> *As I have pointed out on other forums, the obvious, reasonable, expected things that an AGI should and would mandate are COMPLETELY socially unacceptable to just about everyone.*

While insightful in some ways, I also found these comments a bit amusing. First of all, even if true, these points don't obviate the potential value of AGI repair robots. And what about the possibility that an AGI could find politically acceptable technical solutions, which humans haven't been able to conceive given the resources allocated to the problem? Heck, just inventing a cheap material much stronger than those currently used, which could then be deployed to store the waste, would be a big help. I bet this could be done in short order by an AGI scientist with sensors and actuators at the nano-scale, capable of manipulating nano-fibers as readily as we stack blocks or tie knots in rope.

What I saw in the various reactions to my simple, blithe little piece on AGI and nuclear disasters, is the difficulty that even very educated people have in coming to terms with the potential that advanced AGI has. Viewing AGI through the lense of our contemporary limitations, it's all too easy to miss its potential power – just as (to use some far weaker phenomena as analogies) nearly everybody failed to foresee the transformative power of the Internet or mobile phones, even mere years before their advent.

19

Plenty More Room at the Bottom

(Beyond Nanotech to Femtotech)

This H+ Magazine article emerged in 2010, as a result of discussions I was having with my good friend Hugo de Garis. Hugo began his scientific career as a physicist and then verged into AI, working e.g. on novel software/hardware architectures for evolving neural nets in FPGA chips; he has also made a name for himself with outrageous futurist prognostications such as his predictions of a coming "Artilect War" between pro and anti technology forces. But in 2010, after retiring from his position as an AI professor at Xiamen University in China (where he introduced me to my now-wife Ruiting LIan, while I was visiting him there co-running the 2009 AGI Summer School), he returned to his first love of mathematical physics – and began speculating seriously about technologies way smaller than nano. The concept had occurred to me before, but he got me to take it more seriously than I had been, and I put a bit of thought into it, and wrote this article….

I'm pretty sure that advanced AGI – at the human level and significantly beyond – can be achieved using plain ordinary digital computers... For example, server farms like the kind powering Google and Yahoo today. Even if the human brain does make use of some funky quantum nonlocality effects or even weirder physics (as some claim, though there's no evidence of this currently), I really doubt such effects are necessary for achieving humanlike (and transhuman) general intelligence.

But, even if my view on this is right (and the vast majority of physicists and AI researchers, even AGI skeptics, agree with my views on this), that's not to discount the potential for achieving yet greater intelligence by using more exotic computing fabrics. In order to unlock the full depths of computing power implicit in physical reality – in a grain of sand, as Hugo de Garis likes to put it – we're almost surely going to need to go beyond traditional digital computing, and venture into quantum computing and beyond.

We'll need to delve deep into nanotech – and quite possibly beyond, into the world of femtotech, as yet almost wholly uncharted… But in no way ruled out by known physics.

Not long ago, nanotechnology was a fringe topic; now it's a flourishing engineering field, and fairly mainstream. For example, while writing this article, I happened to receive an email advertisement for the "Second World Conference on Nanomedicine and Drug Delivery," in Kerala, India[59]. It wasn't so long ago that nanomedicine seemed merely a flicker in the eyes of Robert Freitas and a few other visionaries[60]!

But every metric system geek knows nano's not so small, really. As you'll recall from the last chapter, a nanometer is 10−9 meters – the scale of atoms and molecules. A water molecule is a bit less than one nanometer long, and a germ is around a thousand nanometers across. On the other hand, a proton has a diameter of a couple femtometers – where a femtometer, at 10−15 meters, makes a nanometer seem positively gargantuan. Now that the viability of nanotech is widely accepted (in spite of some ongoing heated debates about the details), it's time to ask: What about femtotech? Picotech or other technologies at the scales between nano and femto seem relatively uninteresting, because we don't know any basic constituents of matter that exist at those scales. But femtotech, based on engineering structures from subatomic particles, makes perfect conceptual sense, though it's certainly difficult given current technology.

---

59 http://www.nanomedicine.macromol.in/

60 http://www.nanomedicine.com/]

The nanotech field was arguably launched by Richard Feynman's 1959 talk "There's Plenty of Room at the Bottom." As Feynman wrote there,

> *It is a staggeringly small world that is below. In the year 2000, when they look back at this age, they will wonder why it was not until the year 1960 that anybody began seriously to move in this direction.*
>
> ***Why cannot we write the entire 24 volumes of the Encyclopedia Brittanica on the head of a pin?***

But while Feynman's original vision was focused on the nano-scale (the head of a pin, etc.), it wasn't intrinsically restricted to this level. There's plenty of room at the bottom, as he said – and the nano-scale is not the bottom! There's plenty more room down there to explore.

One might argue that, since practical nanotech is still at such an early stage, it's not quite the time to be thinking about femtotech. But technology is advancing faster and faster each year, so it makes sense to think a bit further ahead than contemporary hands-on engineering efforts. Hugo de Garis has been talking to me about femtotech for a while, and has touched on the topic in various lectures and interviews; and he convinced me that the topic is worth looking at in spite of our current lack of knowledge regarding its practical realization. After all, when Feynman gave his "Plenty of Room at the Bottom" lecture, nanotech also appeared radically pie-in-the-sky.

There are many possible routes to femtotech, including some fun ones I won't touch here at all like micro black holes and Bose-Einstein condensation of squarks. I'll focus here largely on a particular class of approaches to femtotech based on the engineering of stable degenerate matter – not because I think this is the only interesting way to think about femtotech, but merely because one has to choose some definite direction to explore if one wants to go into any detail at all.

Physics at the Femto Scale

To understand the issues involved in creating femtotech, you'll first need to recall a few basics about particle physics.

In the picture painted by contemporary physics, everyday objects like houses and people and water are made of molecules, which are made of atoms, which in turn are made of subatomic particles. There are also various subatomic particles that don't form parts of atoms (such as photons, the particles of light, and many others). The behavior of these particles is extremely weird by everyday-life standards – with phenomena like non-local correlations between distant phenomena, observer-dependence of reality, quantum teleportation and lots of other good stuff. But I won't take time here to review quantum mechanics and its associated peculiarities, just to run through a few facts about subatomic particles needed to explain how femtotech might come about.

Subatomic particles fall into two categories: fermions and bosons. These two categories each contain pretty diverse sets of particles, but they're grouped together because they also have some important commonalities.

The particles that serve as the building blocks of matter are all fermions. Atoms are made of protons, neutrons and electrons. Electrons are fermions; and so are quarks, which combine to build protons and neutrons. Quarks appear to occur in nature only in groups, most commonly groups of 2 or 3. A proton contains two up quarks and one down quark, while a neutron consists of one up quark and two down quarks; the quarks are held together in the nucleus by other particles called gluons. Mesons consist of 2 quarks – a quark and an anti-quark. There are six basic types of quark, beguilingly named Up, Down, Bottom, Top, Strange, and Charm. Out of the four forces currently recognized in the universe – electromagnetism, gravity and weak and strong nuclear forces – quarks are most closely associated with the strong nuclear force, which controls most of their dynamics. But quarks also have some interaction with the weak force, e.g. the weak force can cause the transmutation of quarks into different quarks, a phenomenon that underlies some kinds of radioactive decay such as beta decay.

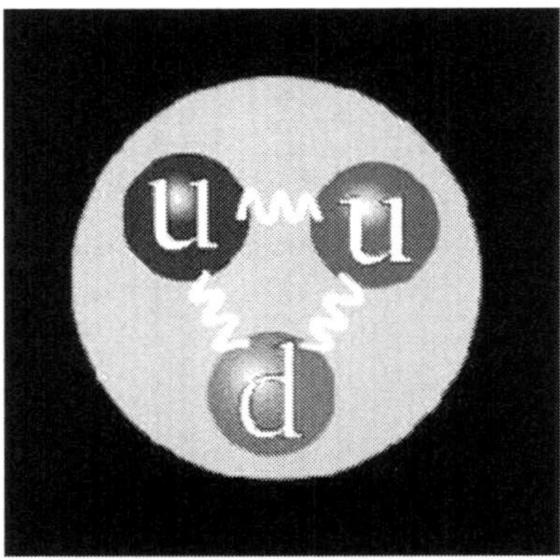

Stylistic depiction of a proton, composed of two Up quarks and one Down quark

On the other hand, bosons are also important – for example photons, the particle-physics version of light, are bosons. Gravitons, the gravity particles proposed by certain theories of gravitation, would also be bosons.

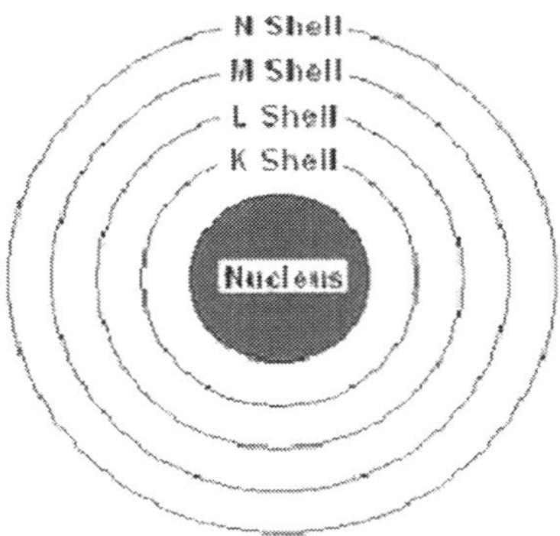

The nucleus of an atom contains protons and neutrons. The electrons are arranged in multiple shells around the nucleus, due

to the Pauli exclusion principle. Also note, this sort of "solar system" model of particles as objects orbiting other objects is just a crude approximation.

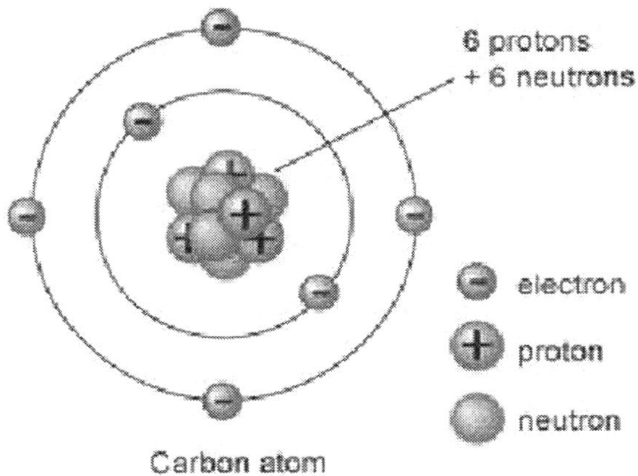

The carbon atom, whose electrons are distributed across two shells.

Finally, just one more piece of background knowledge before we move on to femtotech. Fermions, unlike bosons, obey the **Pauli exclusion principle** – which says that no two identical fermions can occupy the same state at the same time. For example, each electron in an atom is characterized by a unique set of quantum numbers (the principle quantum number which gives its energy level, the magnetic quantum number which gives the direction of orbital angular momentum, and the spin quantum number which gives the direction of its spin). If not for the Pauli exclusion principle, all of the electrons in an atom would pile up in the lowest energy state (the K shell, the innermost shell of electrons orbiting the nucleus of the atom). But the exclusion principle implies that the different electrons must have different quantum states, which results in some of the electrons getting forced to have different positions, leading to the formation of additional shells (in atoms with sufficient electrons).

Degenerate Matter as a Possible Substrate for Femtotech

One can view the Pauli exclusion principle as exerting a sort of "pressure" on matter, which in some cases serves to push particles apart. In ordinary matter this Pauli pressure is minimal compared to other forces. But there is also **degenerate matter** – matter which is so extremely dense that this Pauli pressure or "degeneracy pressure", preventing the constituent particles from occupying identical quantum states, plays a major role. In this situation, pushing two particles close together means that they have effectively identical positions, which means that in order to obey the Pauli exclusion principle, they need to have different energy levels – creating a lot of additional compression force, and causing some very odd states of matter to arise.

For instance, in ordinary matter, temperature is correlated to speed of molecular motion. Heat implies faster motion, and cooling something down makes its component molecules move more slowly. But in degenerate matter, this need not be the case. If one repeatedly cools and compresses a plasma, eventually one reaches a state where it's not possible to compress the plasma any further, because of the exclusion principle that won't let us put two particles in the same state (including the same place). In this kind of super-compressed plasma, the position of a particle is rather precisely defined – but according to a key principle of quantum theory, Heisenberg's uncertainty principle, you can't have accurate knowledge of both the position and the momentum (movement) of a particle at the same time. So the particles in a super-compressed plasma must therefore have highly uncertain momentum – i.e. in effect, they're moving around a lot, even though they may still be very cold. This is just one example of how degenerate matter can violate our usual understanding of how materials work.

At the present time, degenerate matter is mostly discussed in astrophysics, in the context of neutron stars, white dwarf stars, and so forth. It has also been popular in science fiction – for example, in the Star Trek universe, neutronium (matter formed only from large numbers of neutrons, stable at ordinary gravities) is an extremely hard and durable substance, often used as armor, which conventional weapons cannot penetrate or dent at all. But so far neutronium has never been seen in reality. "Strange matter" – defined as matter consisting of an equal number of up, down and strange quarks – is another kind of degenerate matter, with potential applications to femtotech, which I'll return to a little later.

As a substrate for femtotech, degenerate matter appears to have profound potential. It serves as an existence proof that, yes, one can build stuff other than atoms and molecules with subatomic particles. On the other hand, there is the problematic fact that all the currently known examples of degenerate matter exist at extremely high gravities, and derive their stability from this extreme gravitational force. Nobody knows, right now, how to make degenerate matter that remains stable at Earth-level gravities or anywhere near. However, neither has anybody shown that this type of degenerate matter is an impossibility according to our currently assumed physical laws. It remains a very interesting open question.

Bolonkin's Fantastic Femtotech Designs

If you type "femtotech" into a search engine, you'll likely come up with a 2009 paper by A.A. Bolonkin, a former Soviet physicist now living in Brooklyn, entitled "**Femtotechnology: Nuclear Matter with Fantastic Properties**[61]". Equations and calculations notwithstanding, this is an explicitly speculative paper – but the vision it presents is intriguing.

Bolonkin describes a new (and yet unobserved) type of matter he calls "AB-matter", defined as matter which exists at ordinary

---

[61] http://www.scipub.org/fulltext/ajeas/ajeas22501-514.pdf

Earthlike gravities, yet whose dynamics are largely guided by Pauli exclusion principle based degeneracy pressure. He explores the potential of creating threads, bars, rods, tubes, nets and so forth using AB-matter. He argues that

> *"this new 'AB-Matter' has extraordinary properties (for example, tensile strength, stiffness, hardness, critical temperature, superconductivity, supertransparency and zero friction.), which are up to millions of times better than corresponding properties of conventional molecular matter. He shows concepts of design for aircraft, ships, transportation, thermonuclear reactors, constructions and so on from nuclear matter. These vehicles will have unbelievable possibilities (e.g., invisibility, ghost-like penetration through any walls and armor, protection from nuclear bomb explosions and any radiation flux)."*

All this sounds exciting indeed! And the parallels between Bolonkin's diagrams and Drexler's diagrams in *Nanosystems* are obvious. But nowhere in Bolonkin's fascinating thought-piece does he address the million-dollar question of how and why he thinks such structures could be made stable.

I discussed this with Steve Omohundro, a fellow AI researcher and futurist thinker who started his career as a physicist, and Steve very articulately expressed the same "common sense nuclear physics" worries I experienced on reading Bolonkin's paper:

> *A standard model for a nucleus is the "liquid drop" model and it gives pretty good predictions. Basically it treats the nucleus as a liquid with a pretty high surface tension. The nucleons in the center are energetically very happy because they are surrounded by other nucleons attracted by the strong interaction. The nucleons on the surface are not so energetically happy because they interact with fewer other nucleons than they might otherwise. This creates a high effective "surface tension" for the nuclear liquid. That's what*

> *makes nuclei want to be spherical. And when they get too big they become unstable because the surface area is relatively larger and electrostatic repulsion overcomes the nuclear attraction. All of Bolonkin's proposed femtostructures seem unstable to me. His femto rods or whiskers are like streams of water which are subject to instabilities that cause them to break into a sequence of droplets. Imagine one of his rods periodically squeezing inward and outward keeping the volume fixed. If the surface area is decreased the perturbation will be increased and eventually break the rod into droplets.*
>
> *Even if they weren't subject to that instability, there would be tremendous tensile force trying to pull the two ends of a rod together and turning it into a ball (which has a smaller surface area than the same volume cylinder). I didn't see any suggestions for what he wants to use to counteract that tensile force.*

Like me, Steve has a tendency to be open-minded about wild-sounding future possibilities. But open-mindedness must be tempered with a bit of realism.

I'm hoping for a sequel from Bolonkin containing at least back-of-the-envelope stability arguments. But I'm afraid it may not be forthcoming. After a bunch of hunting, in early 2011 I finally managed to get in touch with Alexander Bolonkin and have a brief conversation about femtotech. He turned out to be a really nice guy, and an old friend of my late great friend Valentin Turchin (the Russian dissident, physicist, computer scientist and systems theorist with whom I collaborated on supercompilation in 1999-2001, and to whose memory I dedicated my book *A Cosmist Manifesto*). Like Turchin, Bolonkin had been a Soviet dissident, and he had arrived in New York sometime after Val Turchin, so that Val had played the role of helping him settle into America. He showed me a host of his futurist writings, which echoed many ideas current in the transhumanist community.

Similar to Val Turchin, he had been thinking and writing about these topics (superintelligent robots, cryonics, nanotech, femtotech, immortality on Earth, uploading, etc. etc.) since the 1960s. Not that many modern American techno-futurists realize that the Soviet scientific community got there way before the West! And not only in terms of science-fictional speculation, but also in terms of deep scientific and engineering theory – like femtotechnology!

About femtotech, Bolonkin kindly shared with me some of his thoughts, such as:

> *If we conventionally join the carbon atom to another carbon atom a lot of times, we then get the conventional piece of a coil. If we joint the carbon atom to another carbon atom by the indicated special methods, we then get the very strong single-wall nanotubes, graphene nano-ribbon (super-thin film), armchair, zigzag, chiral, fullerite, torus, nanobud and other forms of nano-materials. That outcome becomes possible because the atomic force (van der Waals force, named for the Dutch physicist Johannes Diderik van der Waals, 1837-1923, etc.) is NON-SPHERICAL and active in the short (one molecule) distance. The nucleon nuclear force also is NON-SPHERICAL and they may also be active about the one nucleon diameter distance. That means we may also produce with them the strings, tubes, films, nets and other geometrical constructions.*
>
> *You can get a stable AB-matter if you put the nucleus in line and calculate the electrostatic (long distance) force which creates the tensile stress and keep this line as string.*

It was evident he'd done a lot of back-of-the-envelope calculations beyond what he'd written in his one brief paper on the topic – and I also got the feeling he wasn't feeling any great urgency to write up the details in any formal way, for others to peruse and continue his work. I can well understand this, since I myself have probably written up only 25% or so of the interesting technical ideas I've worked out in various scribblings and notes here and there. Writing stuff up just isn't nearly as fun as figuring it out in the first place, and often takes more time – time that could be spent figuring out more and more cool stuff! It would be great for some young transhumanist physicist to spend a few months carefully picking his brain on the topic, and maybe write a joint opus on the theory of femtotech. If I weren't so busy pushing toward AGI, I'd be sorely tempted to spend a few months that way myself!

Might Dynamic Stabilization Work on Degenerate Matter?

And in spite of his initial skeptical reaction, after a little more thought Steve had a rather interesting brainstorm:

> *I just had a thought about how to stabilize degenerate femtomatter: use dynamic stabilization. The classic example is the shaking inverted pendulum. An upside down pendulum is unstable, falling either left or right if perturbed. But if you shake the base at a sufficiently high frequency, it adds a "pondermotive"*
>
> *pseudopotential which stabilizes the unstable fixed point. Here's a video of a guy who built one*[62]
>
> *The same approach can stabilize fluid instabilities. If you turn a cup of fluid upside down, the perfectly flat surface is an unstable equilibrium. The Rayleigh-Taylor instability causes ripples to grow and the fluid to spill out. But, I remember seeing a guy years ago who put a cup of oil in a shaking apparatus and was able to turn it upside down without it spilling. So the oscillations were able to stabilize all the fluid modes at once. I wonder if something similar might be used to stabilize degenerate matter at the femto scale?*

---

[62] http://www.youtube.com/watch?v=rwGAzy0noU0

A fascinating idea indeed! Instead of massive gravity or massive heat, perhaps one could use incredibly fast, low-amplitude vibrations to stabilize degenerate matter. How to vibrate subatomic particles that fast is a whole other matter, and surely a difficult engineering problem – but still, this seems a quite promising avenue. It would be interesting to do some mathematics regarding the potential dynamic stabilization of various configurations of subatomic particles subjected to appropriate vibrations.

An inverted pendulum kept vertical via dynamic stabilization. The rod would rotate and fall down to one side or another if it weren't vibrating. But if it's vibrated very fast with low amplitude, it will remain upright due to dynamic stabilization. Con *Conceivably a similar phenomenon could be used to make stable degenerate matter, using very fast femtoscale vibrations.*

Of course, such metaphorical ideas must be taken with a grain of salt. When I think about the "liquid drop" model of the nucleus, I'm somewhat reminded of how the genius inventor Nikola Tesla intuitively modeled electricity as a fluid. This got him a long way compared to his contemporaries, leading him to develop AC power and ball lightning generators and all sorts of other amazing stuff – yet it also led to some mistakes, and caused him to miss some things that are implicit in the mathematics of electromagnetism but not in the intuitive metaphorical "electricity as fluid" model. For instance Tesla's approach to wireless power transmission was clearly misguided in some respects (even if it did contain some insights that haven't yet been fully appreciated), and this may have been largely because of the limitations of his preferred fluid-dynamics metaphor for electricity. Where degenerate matter is concerned, metaphors to liquid drops and macroscopic shaking apparatuses may be very helpful for inspiring additional experiments, but eventually we can expect rigorous theory to far outgrow them.

The bottom line is, in the current state of physics, nobody can analytically solve the equations of nuclear physics except in special simplified cases. Physicists often rely on large-scale computer simulations to solve the equations in additional cases – but these depend on various technical simplifying assumptions, which are sometimes tuned based on conceptual assumptions about how the physics works. Intuitive models like "nucleus as water droplet" are based on the limited set of cases in which we've explored the solutions of the relevant equations using analytical calculations or computer simulations. So, based on the current state of the physics literature, we really don't know if it's possible to build stable structures of the sort Bolonkin envisions. But there are surely worthwhile avenues to explore, including Steve's intriguing suggestion.

Gell-Mann Gives Femtotech A Definite Maybe

A few weeks ago, while at an event in San Francisco, I was thrilled to have the opportunity to discuss femtotech with Murray Gell-Mann – who is not only a Nobel Prize winning physicist, but also one of the world's ultimate gurus on quarks, since he invented and named the concept and worked out a lot of the theory of their behavior. I knew my friend Hugo de Garis had briefly discussed femtotech with Gell-Mann a decade and a half previously, but that he hadn't expressed any particular thoughts on the topic. I was curious if Gell-Mann's views on the topic had perhaps progressed a bit.

To my mild disappointment, Gell-Mann's first statement to me about femtotech was that he had never thought about the topic seriously. However he went on to say that it seemed to be a reasonable idea to pursue. As a mathematician and AI guy dabbling in physics, this was a relief to me– at least the great physicist didn't laugh at me!!

When I probed Gell-Mann about degenerate matter, he spent a while musing about the possible varieties of degenerate matter in which the ordinary notion of quark confinement is weakened. “Confinement” is the property that says quarks cannot be isolated singularly, and therefore cannot be directly observed, but can only be observed as parts of other particles like protons and neutrons. At first it was thought that quarks can only be observed in triplets, but more recent research suggests the possibility of “weak confinement” that lets you observe various aspects of individual quarks in an isolated way. Quark-gluon plasmas[63], which have been created in particle accelerators using very high temperatures (like, 4 trillion degrees!), are one much-discussed way of producing “almost unconfined” quarks. But Gell-Mann felt the possibilities go far beyond quark-gluon plasmas. He said he thought it possible that large groups of quarks could potentially be weakly-confined in more complex ways, that nobody now understands.

So after some fun discussion in this vein, I pressed Gell-Mann specifically on whether understanding these alternative forms of weak multi-quark confinement might be one way to figure out how to build stable degenerate matter at Earth gravity…

His answer was, basically, definitely maybe.

Then we changed the topic to AI and the Singularity, where I'm on firmer ground – and there he was a little more positive, actually. He said he thought it was crazy to try to place a precise

---

[63] http://www.bnl.gov/rhic/physics.asp

date on the Singularity, or to estimate anything in detail about it in advance... but he was sympathetic to the notion of accelerating technological change, and very open to the idea that massively more change is on the way. And, contra his fellow physicist Roger Penrose, he expressed doubt that quantum computing (let alone femtocomputing) would be necessary for achieving human-level AI. Even if the human brain somehow uses strange quantum effects in some particulars, he felt, digital computers should most likely be enough to achieve human-level intelligence.

A few moments later at the same event, I asked a young Caltech physics postdoc the same questions about degenerate matter and femtotech – and he gave a similar answer, only mildly more negative in tone. He said it seemed somewhat unlikely that one could make room-temperature stable structures using degenerate matter, but that he couldn't think of any strong reason why it would be impossible....

Currently, it seems, where degenerate matter based femtotech is concerned – nobody knows.

Strange Matter and Other Strange Matters

Gell-Mann's comments reminded me of strangelets[64] – strange hypothetical constructs I first found out about a few years ago when reading about some strange people who had the strange idea that the Large Hadron Collider might destroy the world[65] by unleashing a strange chain reaction turning the Earth into strangelets. Fortunately this didn't happen – and it seems at least plausible that strangelets might pose a route to stable degenerate matter of a form useful for femtotech.

---

[64] http://www.wisegeek.com/what-is-a-strangelet.htm
[65] http://public.web.cern.ch/public/en/lhc/safety-en.html

A strangelet is (or would be, if they exist at all, which is unknown) an entity consisting of roughly equal numbers of up, down and strange quarks. A small strangelet would be a few femtometers across, with around the mass of a light nucleus. A large strangelet could be meters across or more, and would then be called a "strange star" or a "quark star."

In a (hypothetical) strange star, quarks are not confined in the traditional sense, but may still be thought of "weakly confined" in some sense (at least that's Gell-Mann's view)

So far, all the known particles with strange quarks – like the Lambda particle – are unstable. But there's no reason to believe that states with a larger number of quarks would have to suffer from this instability. According to Bodmer[66] and Witten's[67] "strange matter hypothesis," if enough quarks are collected together, you may find that the lowest energy state of the collective is a strangelet, i.e. a state in which up, down, and strange quarks are roughly equal in number.

---

[66] http://prd.aps.org/abstract/PRD/v4/i6/p1601_1

[67] http://prd.aps.org/abstract/PRD/v30/i2/p272_1

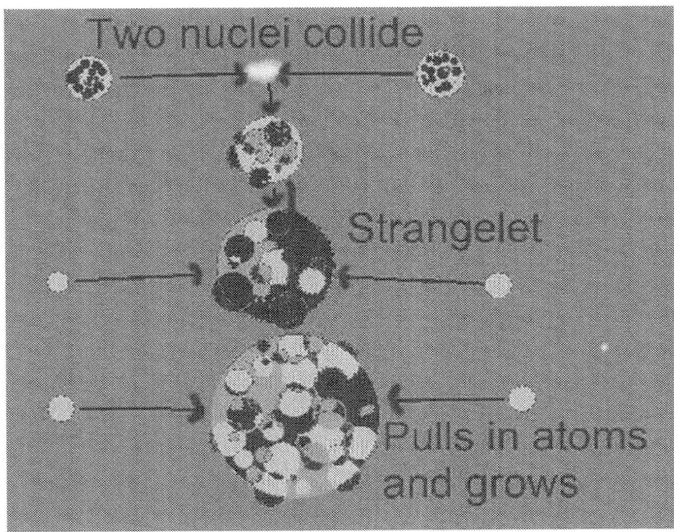

The hypothetical chain reaction via which strangelets eat the Earth

So, where does the End of the World come in? There are some interesting (albeit somewhat speculative) arguments to the effect that if a strangelet encounters ordinary matter, it could trigger a chain reaction in which the ordinary matter gets turned into strangelets, atom by atom at an accelerating pace. Once one strangelet hits a nucleus, it would likely turn it into strange matter, thus producing a larger and more stable strangelet, which would in turn hit another nucleus, etc. Goodbye Earth, hello huge hot ball of strange matter. This was the source of the worries about the LHC, which did not eventuate since when the LHC was utilized no strangelets were noticeably produced.

One of the many unknowns about strangelets is their surface tension – nobody knows how to calculate this, at present. If strangelets' surface tension is strong enough, large stable strangelets should be possible – and potentially, strangelets with complex structure as femtotech requires.

And of course, nobody knows what happens if you vibrate strangelets very very fast with small amplitude – can you produce stable strangelets via dynamic stabilization? Could this be a path to viable femtotechnology, even if stable strangelets don't occur in nature. After all, carbon nanotubes appear not to occur in nature either.

The Future of Femtotech

So what's the bottom line – is there still more room at the bottom?

Nanotech is difficult engineering based on mostly known physics. Femtotech, on the other hand, pushes at the boundaries of known physics. When exploring possible routes to femtotech, one quickly runs up against cases where physicists just don't know the answer.

Degenerate matter of one form or another seems a promising potential route to femtotech. Bolonkin's speculations are intriguing, as are the possibilities of strangelets or novel weakly confined multi-quark systems. But the issue of stability is a serious one; nobody yet knows whether large strangelets can be made stable, or whether degenerate matter can be created at normal gravities, nor whether weakly confined quarks can be observed at normal temperatures, etc. Even where the relevant physics equations are believed known, the calculations are too hard to do given our present analytical and computational tools. And in some cases, e.g. strangelets, we run into situations where different physics theories held by respected physicists probably yield different answers.

Putting my AI futurist hat on for a moment, I'm struck by what a wonderful example we have here of the potential for an only slightly superhuman AI to blast way past humanity in science and engineering. The human race seems on the verge of understanding particle physics well enough to analyze possible routes to femtotech. If a slightly superhuman AI, with a talent for physics, were to make a few small breakthroughs in computational physics, then it might (for instance) figure out how to make stable structures from degenerate matter at Earth gravity. Bolonkin-style femtostructures might then become plausible, resulting in femtocomputing – and the slightly superhuman AI would then have a computational infrastructure capable of supporting massively superhuman AI. Can you say "Singularity"? Of course, femtotech may be totally unnecessary in order for a Vingean Singularity to occur (in fact I strongly suspect so). But be that as it may, it's interesting to think about just how much practical technological innovation might ensue from a relatively minor improvement in our understanding of fundamental physics.

Is it worth thinking about femtotech now, when the topic is wrapped up with so much unresolved physics? I think it is, if for no other reason than to give the physicists a nudge in certain directions that might otherwise be neglected. Most particle physics work – even experimental work with particle accelerators – seems to be motivated mainly by abstract theoretical interest. And there's nothing wrong with this – understanding the world is a laudable aim in itself; and furthermore, over the course of history, scientists aiming to understand the world have spawned an awful lot of practically useful by-products. But it's interesting to realize that there are potentially huge practical implications waiting in the wings, once particle physics advances a little more – if it advances in the right directions.

So, hey, all you particle physicists and physics funding agency program managers reading this article (and grumbling at my oversimplifications: sorry, this is tough stuff to write about for a nontechnical audience!), please take note – why not focus some attention on exploring the possibility of complexly structured degenerate matter under Earthly conditions, and other possibly femtotech-related phenomena?

Is there still plenty more room at the bottom, after the nanoscale is fully explored – for the construction of super-advanced AGI and other future technologies? It seems quite possibly so – but we need to understand what goes on way down there a bit better, before we can build stuff at the femtoscale. Fortunately, given the exponentially accelerating progress we're seeing in some relevant areas of technology, the wait for this understanding and the ensuing technologies may not be all that long.

20

The Singularity Institute's Scary Idea (and Why I Don't Buy It)

This essay, posted on my blog "The Multiverse According to Ben" in 2009, attracted more comments than any other blog post I've made[68] *– because the organization it discussed, SIAI, possessed a great number of passionate supporters, as well as a great number of passionate detractors…. The version given here was edited by me at some later point, but is essentially the same as the blog post version….*

I felt some reluctance about including this essay here in this book, because SIAI has somewhat moved on from where it was back in 2009 – including a name change (to MIRI, Machine Intelligence Research Institute) and a bunch of new management. However, in the end I decided to include it, because in spite of MIRI's greater professionalism, the basic SIAI perspective I critiqued in my essay is still there at the heart of MIRI; and in various ways the debate that this essay represents is still ongoing. As I write these words in 2014, Oxford philosopher Nick Bostrom's book ***Superintelligence*** *has recently been published – and in large part, it consists of a more academic and sober presentation of a minor variant of the line of thinking I call here "The Singularity Institute's Scary Idea" ….*

In a 2012 online dialogue with new MIRI executive Luke Muehlhauser, I delved into these same issues in an updated way, including responses to some of Bostrom's essays which form the conceptual core of his ***Superintelligence*** *book. That*

---

68 http://multiverseaccordingtoben.blogspot.hk/2010/10/singularity-institutes-scary-idea-and.html

dialogue is contained in my book ***Between Ape and Artilect****, and can fairly be considered the next step in the debate after the essay you're about to read:*

This essay focuses on some of the ideas of the organization that until 2013 was called SIAI, the Singularity Institute for Artificial Intelligence (and is now called MIRI, the Machine Intelligence Research Institite). This is a group that, for all its flaws (some of which I'll focus on here), is one of the more ambitious and interestingly-focused organizations on the planet.

I have pretty severe issues with a particular line of thinking this group has advocated, that I'll refer to here as "SIAI's Scary Idea."

Roughly, the Scary Idea posits that: *If I or anybody else actively trying to build advanced AGI succeeds, we're highly likely to cause an involuntary end to the human race.*

When SIAI rebranded as MIRI, they changed their marketing focus and organizational structure significantly – but at core, they certainly didn't drop the Scary Idea.

Ah, and one more note to forestall (or maybe provoke, for some readers) confusion: the Singularity Institute for AI is not, and never was, affiliated with the other California futurist organization called Singularity University ... though there are some overlaps. Ray Kurzweil founded SU and advised SIAI; and I previously advised SIAI and currently advise SU. Further, in 2013, as part of rebranding itself as MIRI, SIAI sold the Singularity Summit conference series to Singularity University. Small Singularitarian world!!

As the organization was called SIAI for most of the time I interacted with them, and when I wrote the initial version of this chapter (as a blog post), and as "SIAI/MIRI" looks ugly, I'm going to keep referring to "SIAI" throughout this chapter.

Since this chapter is going to be somewhat critical, I first want to clarify that I'm not in disagreement with the existence of SIAI as an institution, nor with the majority of their activities. I only disagree with certain positions habitually held by some SIAI researchers, and by the community of individuals heavily involved with SIAI.

Actually, SIAI has been quite good to me, overall. I enjoyed all the Singularity Summits, which they've hosted, very much (well, I didn't attend the 2011 and 2012 Summits, but I spoke at all the previous ones). I feel these Summits have played a major role in the advancement of society's thinking about the future, and I've felt privileged to speak at them. I applaud SIAI for having consistently being open to Summit speakers whose views are strongly divergent from those commonly held by the SIAI community.

Additionally, to their credit, in 2008, SIAI and my company Novamente LLC seed-funded the OpenCog[69] open-source AGI project, which was based on software code spun out from Novamente. The SIAI/OpenCog relationship diminished substantially when Tyler Emerson passed the leadership of SIAI along to Michael Vassar, but it was very instrumental in getting OpenCog off the ground. I've also enjoyed working with Michael Vassar on the Board of Humanity+, of which I am Chair and he is a Board member.

When SIAI was helping fund OpenCog, I took the title of "Director of Research" of SIAI, but I never actually directed any research there apart from OpenCog. The other SIAI research was always directed by others, which was fine with me. There were occasional discussions about operating in a more unified manner, but it didn't happen. All this is perfectly ordinary in a small start-up type organization. And once SIAI decided OpenCog was no longer within its focus, after a bit of delay I

---

[69] http://opencog.org/

decided it didn't make sense for me to hold the Director of Research title anymore, since as things were evolving, I wasn't directing any SIAI research. I remain as an Advisor to SIAI, which is going great. Given all this history, I feel a bit like by writing this chapter I'm "biting the hand that once fed me" - or at least, that once fed OpenCog, in its infancy. But, one of the positive attributes of SIAI is the emphasis it places on rationality and devotion to truth. So in this sense, a chapter like this, which rationally questions the truth of some of SIAI's frequent assertions, is very much in the SIAI spirit.

And so, without further ado...

SIAI's Scary Idea (Which I Don't Agree With)

The collection of concepts I call "SIAI's Scary Idea" has been worded in many different ways by many different people. In the following paragraph I've tried to word it in a way that captures the idea fairly if approximately, and won't annoy too many people:

SIAI's Scary Idea: Progressing toward advanced AGI without a design for "provably non-dangerous AGI" (or something closely analogous, often called "Friendly AI" in SIAI lingo) is highly likely to lead to an involuntary end for the human race.

One of the issues with the Scary Idea is that it's rarely clarified what "provably" really means. A mathematical proof can only be applied to the real world in the context of some assumptions, so one reasonable interpretation of "a provably non-dangerous AGI" is "an AGI whose safety is implied by mathematical arguments together with assumptions that are believed reasonable by some responsible party?" Of course, this begs the question of who responsible party is. Perhaps "the overwhelming majority of scientists," or some self-appointed group of experts, such as SIAI itself?

If you read the preceding chapters of this book you know that, although I don't agree with the Scary Idea, I do agree that the development of advanced AGI has significant risks associated with it. There are also dramatic potential benefits associated with it, including the potential of protection against risks from other technologies (like nanotech, biotech, narrow AI, etc.). So the development of AGI has difficult cost-benefit balances associated with it, just like the development of many other technologies.

I also agree with Nick Bostrom, a host of Science Fiction writers, and many others, that AGI is a potential "existential risk." Meaning that, in the worst case, AGI could wipe out humanity entirely. I think nanotech, biotech, and narrow AI could also do so, along with a bunch of other things.

I certainly don't want to see the human race wiped out! I personally would like to transcend the legacy human condition and become a transhuman superbeing. And I would like everyone else to have the chance to do so, if they want to. However, even though I think this kind of transcendence will be possible, and will be desirable to many, I wouldn't like to see anyone forced to transcend in this way. I would like to see the good old fashioned human race continue, if there are humans who want to maintain their good old fashioned humanity, even if other options are available.

But SIAI's Scary Idea goes way beyond the mere statement that there are risks as well as benefits associated with advanced AGI, and that AGI is a potential existential risk.

Finally, I note that most of the other knowledgeable futurist scientists and philosophers who have come into close contact with SIAI's perspective, also don't accept the Scary Idea. Examples include Robin Hanson and Ray Kurzweil.

Obviously, I'm not opposed to anyone having radical ideas that their respected peers mostly don't accept! I totally get that. My own approach to AGI is somewhat radical, and most of my friends in the AGI research community, while they respect my work and see its potential, aren't quite as enthused about it as I am. Radical positive changes are often brought about by people who clearly understand certain radical ideas well before anyone else "sees the light." However, my own radical ideas are not telling whole research fields that if they succeed they're bound to kill everybody. So, in that way, they are in a much different situation than the Scary Idea!

What is the Argument for the Scary Idea?

Although an intense interest in rationalism is one of the hallmarks of the SIAI community, I still have not yet seen a clear logical argument for the Scary Idea laid out anywhere. SIAI Media Director Michael Anissimov says he's working on a book that will present such an argument, along with other topics. But for the moment, if one wants a clear argument for the Scary Idea, one basically has to construct it oneself.

As far as I can tell from discussions and the available online material, some main ingredients of peoples' reasons for believing the Scary Idea are ideas like:

1. If one pulled a random mind from the space of all possible minds, the odds of it being friendly to humans (as opposed to, e.g., utterly ignoring us, and being willing to repurpose our molecules for its own ends) are very low.
2. Human value is fragile[70], as well as complex, so if you create an AGI with a roughly-human-like value system, then this may not be good enough, and it is likely to rapidly diverge into something with little or no respect for human values.
3. "Hard takeoffs" (in which AGIs recursively self-improve and massively increase their intelligence) are fairly likely

---

[70] http://lesswrong.com/lw/y3/value_is_fragile/

once AGI reaches a certain level of intelligence; and humans will have little hope of stopping these events.

4. A hard takeoff, unless it starts from an AGI designed in a "provably Friendly" way, is highly likely to lead to an AGI system that doesn't respect the rights of humans to exist.

Note, I'm not directly quoting any particular thinker associated with SIAI here. I'm merely summarizing, in my own words, ideas that I've heard and read very often from various individuals associated with SIAI.

If you put the above points all together, you come up with a heuristic argument for the Scary Idea. Roughly, the argument goes something like: *If someone builds an advanced AGI without a provably Friendly architecture, probably it will have a hard takeoff, and then probably this will lead to a superhuman AGI system with an architecture drawn from the vast majority of mind-architectures that are not sufficiently harmonious with the complex, fragile human value system to make humans happy and keep humans around.*

The line of argument makes sense, if you accept the premises. But, I don't. I think the first of the above points is reasonably plausible, though I'm not by any means convinced. I think the relation between breadth of intelligence and depth of empathy is a subtle issue which none of us fully understands (yet). It's possible that with sufficient real-world intelligence tends to come a sense of connectedness with the universe that militates against squashing other sentients.

But I'm not terribly certain of this, any more than I'm terribly certain of its opposite. I agree much less with the final three points listed above, and I haven't seen any careful logical arguments for these points. I doubt human value is particularly fragile. Human value has evolved and morphed over time and will continue to do so. It already takes multiple different forms. It will likely evolve in future in coordination with AGI and other technology. I think it's fairly robust.

I think a hard takeoff is possible, though I don't know how to estimate the odds of one occurring with any high confidence. I think it's very unlikely to occur until we have an AGI system that has very obviously demonstrated general intelligence at the level of a highly intelligent human. And I think the path to this "hard takeoff enabling" level of general intelligence is going to be somewhat gradual, not extremely sudden.

I don't have any strong sense of the probability of a hard takeoff, from an apparently but not provably human-friendly AGI, leading to an outcome likable to humans. I suspect this probability depends on many features of the AGI, which we will identify over the next years and decades, via theorizing based on the results of experimentation with early-stage AGIs.

Yes, you may argue: the Scary Idea hasn't been rigorously shown to be true... but what if it IS true?

OK but ... pointing out that something scary is *possible*, is a very different thing from having an argument that it's *likely*.

The Scary Idea is certainly something to keep in mind, but there are also many other risks to keep in mind, some much more definite and palpable. Personally, I'm a lot more worried about nasty humans taking early-stage AGIs and using them for massive destruction, than about speculative risks associated with little-understood events like hard takeoffs.

Is Provably Safe or "Friendly" AGI A Feasible Idea?

The Scary Idea posits that if someone creates advanced AGI that isn't somehow provably safe, it's almost sure to kill us all.

But not only am I unconvinced of this, I'm also quite unconvinced that a "provably safe" AGI is even feasible. The idea of provably safe AGI is typically presented as something that would exist within mathematical computation theory or some variant thereof. So, right off the bat, that's one obvious limitation of the idea, as mathematical computers don't exist in the real world, and real-world physical computers must be interpreted in terms of the laws of physics, and humans' best understanding of the "laws" of physics seem to radically change from time to time. So, even if there were a design for provably safe real-world AGI, based on current physics, the relevance of the proof might go out the window when physics next gets revised.

Also, there are always possibilities like: the alien race that is watching us and waiting for us to achieve an IQ of 333, at which point it will swoop down upon us and eat us, or merge with us. We can't rule this out via any formal proof, and we can't meaningfully estimate the odds of it either. Yes, this sounds science-fictional and outlandish; but is it really more outlandish and speculative than the Scary Idea?

A possibility that strikes me as highly likely is that, once we have created advanced AGI, and have linked our brains with it collectively, most of our old legacy human ideas (including physical law, aliens, and Friendly AI) will seem extremely limited and ridiculous.

Another issue is that the goal of "Friendliness to humans" or "safety" or whatever you want to call it, is rather nebulous and difficult to pin down. Science fiction has explored this theme extensively. So, even if we could prove something about "smart AGI systems with a certain architecture that are guaranteed to achieve goal G," it might be infeasible to apply this to make AGI systems that are safe in the real-world, simply because we don't know how to boil down the everyday intuitive notions of "safety" or "friendliness" into a mathematically precise goal G like the proof refers to.

This is related to the point Eliezer Yudkowsky makes that "'value' is complex." Actually, human value is not only complex, it's nebulous and fuzzy, and ever-shifting, and humans largely grok it by implicit procedural, empathic and episodic knowledge, rather than explicit declarative or linguistic knowledge. Transmitting human values to an AGI is likely to be best done via interacting with the AGI in real life, but this is not the sort of process that readily lends itself to guarantees or formalization.

But, setting those worries aside, is the computation-theoretic version of provably safe AI even possible? Could one design an AGI system and prove, in advance, that, given certain reasonable assumptions about physics and its environment, it would never veer too far from its initial goal (e.g. a formalized version of the goal of treating humans safely, or whatever)?

I very much doubt one can do so, except via designing a fictitious AGI that can't really be implemented because it uses an infeasible amount of computational resources. In fact, I've put a fair amount of thought into an AGI design like this, which I call GOLEM. I'll briefly describe GOLEM in a later chapter, because I think it's an interesting intellectual experiment. However, suspect it's too computationally wasteful to be practically feasible, at least for now (pre-Singularity).

I strongly suspect that to achieve high levels of general intelligence using realistically limited computational resources, one is going to need to build systems with a nontrivial degree of *fundamental unpredictability* to them. This is what neuroscience suggests, it's what my concrete AGI design work suggests, and it's what my theoretical work on GOLEM and related ideas suggests. And none of the public output of SIAI researchers or enthusiasts has given me any reason to believe otherwise, yet.

Practical Implications

The above discussion of SIAI's Scary Idea may just sound like fun science-fictional speculation, but this chapter originated out of a frustrating practical situation that arose as a direct result of the Scary Idea. In 2010 I wrote an entry on my blog about the OpenCog project, and the comments field on the blog post got totally swamped with SIAI-influenced people saying stuff in the vein of: *Creating an AGI without a proof of Friendliness is essentially equivalent to killing all people! So I really hope your OpenCog work fails, so you don't kill everybody!!!*

On the most extreme side, one amusing/alarming quote from a commentator (probably not someone directly affiliated with SIAI) was "if you go ahead with an AGI when you're not 100% sure that it's safe, you're committing the Holocaust." There were many other comments in the same rough vein, and also a number of other similar comments directed to me privately via email.

If one fully accepts SIAI's Scary Idea, then one should not work on practical AGI projects, ever. Nor should one publish papers on the theory of how to build AGI systems. Instead, one should spend one's time trying to figure out an AGI design that is somehow provable-in-advance to be a Nice Guy. For this reason, SIAI's research group is not currently trying to do any practical AGI work.

Actually, so far as I know, my "GOLEM" AGI design (mentioned above) is closer to a "provably Friendly AI" than anything the SIAI research team has come up with. (At least anything they have made public.) I definitely agree that AGI ethics is a very important problem, but I doubt the problem is most effectively addressed by theory alone. I think the way to achieve to a useful real-world understanding of AGI ethics is going to be to

- Build some early-stage AGI systems, e.g. artificial toddlers, scientists' helpers, video game characters, robot maids and butlers, etc.
- Study these early-stage AGI systems empirically, with a focus on their ethics as well as their cognition.

- Attempt to arrive at a solid theory of AGI intelligence and ethics based on a combination of conceptual and experimental-data considerations.
- Have humanity collectively plot the next steps from there. Based on the theory we find, maybe we go ahead and create a superhuman AI capable of hard takeoff, or maybe we pause AGI development because of the risks. Maybe we build an "AGI Nanny" to watch over the human race and prevent AGI or other technologies from going awry. Whatever choice we make then, it will be made based on far better knowledge than we have right now.

So what's wrong with this approach? Nothing, really. If you hold the views of most AI researchers or futurists. There are plenty of disagreements about the right path to AGI, but wide and implicit agreement that something like the above path is sensible.

But, if you adhere to SIAI's Scary Idea, there's a big problem with this approach – because, according to the Scary Idea, there's too huge of a risk that these early-stage AGI systems are going to experience a hard takeoff and self-modify into something that will destroy us all. But I just don't buy the Scary Idea.

I do see a real risk that, if we proceed in the manner I'm advocating, some nasty people will take the early-stage AGIs and either use them for bad ends, or proceed to hastily create a superhuman AGI that then does bad things of its own volition. These are real risks that must be thought about hard, and protected against as necessary, but they are different from the Scary Idea, and not really so different from the risks implicit in a host of other advanced technologies.

Conclusion

Although I think SIAI is performing a useful service by helping bring these sorts of ideas to the attention of the futurist community (alongside the other services they're performing, like the wonderful Singularity Summits), that all said, I think the Scary Idea is potentially a harmful one. At least, it *would* be a harmful one, if more people believed it. Luckily, such a viewpoint is currently restricted to a rather small subset of the futurist community.

Many people die each day, and many others are miserable for various reasons -- and all sorts of other advanced and potentially dangerous technologies are currently under active development. My own view is that unaided human minds may well be unable to deal with the complexity and risk of the world that human technology is unleashing. I actually suspect that our best hope for survival and growth through the 21st century is to create advanced AGIs to help us on our way: To cure disease, develop nanotech and better AGI, invent new technologies, and help us keep nasty people from doing destructive things with advanced technology.

I think that to avoid actively developing AGI, out of speculative concerns like the Scary Idea, would be an extremely bad idea.

That is, rather than "if you go ahead with an AGI when you're not 100% sure that it's safe, you're committing the Holocaust," I suppose my view is closer to "if you avoid creating beneficial AGI because of speculative concerns, then you're killing my grandma!" (Because advanced AGI will surely be able to help us cure human diseases and vastly extend and improve human life.)

So perhaps I could adopt the slogan: "**You don't have to kill my grandma to avoid the Holocaust!"**... But really, folks. Well, you get the point.

Humanity is on a risky course altogether, but no matter what I decide to do with my life and career (and no matter what Bill Joy or Jaron Lanier or Bill McKibben, etc., write), the race is not going to voluntarily halt technological progress. It's just not happening.

We just need to accept the risk, embrace the thrill of the amazing time we were born into, and try our best to develop near-inevitable technologies like AGI in a responsible and ethical way.

To me, responsible AGI development doesn't mean fixating on speculative possible dangers and halting development until ill-defined, likely-unsolvable theoretical/philosophical issues are worked out to everybody's (or some elite group's) satisfaction.

Rather, it means proceeding with the work carefully and openly, learning what we can as we move along, and letting experiment and theory grow together, as they have been doing quite successfully for the last few centuries, at a fantastically accelerating pace.

21

Does Humanity Need an AI Nanny?

This H+ Magazine article, from 2011, is one of my lines of thought that I feel most ambivalent about. It's a case where my logical mind pushes in one direction, and my emotions in another. Emotionally, I'm a freedom-loving American anarchist freak by nature, and I hate the idea of being constantly observed and monitored by critical eyes, especially eyes connected to actuators with the power to stop me from doing what I want. Logically, though, it seems to me that given the increasing dangers coming along with increasingly advanced technology, universal surveillance or sousveillance are probably the only rational courses, if we want to survive long enough to transcend to the next stages of intelligence evolution….

The article attracted a bit of attention, including getting me invited to participate in an episode of Michio Kaku's popular science TV show, for which I was filmed talking about the AI Nanny in the office of a company doing video surveillance of various offices – sitting in front of a bunch of monitors showing the outputs of security cameras. Not my usual work environment! …. But meeting Kaku was cool; and we discussed string theory and various more out-there physics ideas, on the van ride to the filming location.

One possible solution to the acute risks posed by rapidly advancing technology development is to build a powerful yet limited AGI system, with the explicit goal of keeping things on the planet under control while we figure out the hard problem of how to create a probably positive Singularity. That is: to create an "AI Nanny."

The AI Nanny would forestall a full-on Singularity for a while, restraining it into what Max More has called a Surge[71], and giving us time to figure out what kind of Singularity we really want to build and how. It's not entirely clear that creating such an AI Nanny is plausible, but I've come to the conclusion it probably is. Whether or not we should *try* to create it, though – that's a different question.

The AI Nanny in Detail

Let me be more precise about what kind of "Nanny" I'm thinking of.... Imagine an advanced AGI system with

- General intelligence somewhat above the human level, but not too dramatically so – maybe, qualitatively speaking, as far above humans as humans are above apes
- Interconnection to powerful worldwide surveillance systems, online and in the physical world
- Control of a massive contingent of robots (e.g. service robots, teacher robots, etc.) and connectivity to the world's home and building automation systems, robot factories, self-driving cars, and so on and so forth
- A cognitive architecture featuring an explicit set of goals, and an action selection system that causes it to choose those actions that it rationally calculates will best help it achieve those goals
- A set of preprogrammed goals including the following aspects:
- A strong inhibition against modifying its preprogrammed goals

71 http://strategicphilosophy.blogspot.com/2009/06/how-fast-will-future-arrive-how-will.html

- A strong inhibition against rapidly modifying its general intelligence
- A mandate to cede control of the world to a more intelligent AI within 200 years
- A mandate to help abolish human disease, involuntary human death, and the practical scarcity of common humanly-useful resources like food, water, housing, computers, etc.
- A mandate to prevent the development of technologies that would threaten its ability to carry out its other goals
- A strong inhibition against carrying out actions with a result that a strong majority of humans would oppose, if they knew about the action in advance

A mandate to be open-minded toward suggestions by intelligent, thoughtful humans about the possibility that it may be misinterpreting its initial, preprogrammed goals

There you have it: an "AI Nanny."

Obviously, this sketch of the AI Nanny idea is highly simplified and idealized – a real-world AI Nanny would have all sort of properties not described here, and might be missing some of the above features, substituting them with other related things. My point here is not to sketch a specific design or requirements specification for an AI Nanny, but rather to indicate a fairly general class of systems that humanity might build.

The nanny metaphor is chosen carefully. A nanny watches over children while they grow up, and then goes away. Similarly, the AI Nanny would not be intended to rule humanity on a permanent basis – only to provide protection and oversight while we "grow up" collectively; to give us a little breathing room so we can figure out how best to create a desirable sort of Singularity.

A large part of my personality rebels against the whole AI Nanny approach – I'm a rebel and a nonconformist; I hate bosses and bureaucracies and anything else that restricts my freedom. But, I'm not a political anarchist – because I have a strong suspicion that if governments were removed, the world would become a lot worse off, dominated by gangs of armed thugs imposing even less pleasant forms of control than those exercised by the US Army and the CCP and so forth. I'm sure government could be done a lot better than any country currently does it – but I don't doubt the need for some kind of government, given the realities of human nature. Well, maybe – *maybe* – the need for an AI Nanny falls into the same broad category. Perhaps, like government, an AI Nanny is a relatively offensive thing, that is nonetheless a practical necessity due to the unsavory aspects of human nature.

We didn't need government during the Stone Age – because there weren't that many of us, and we didn't have so many dangerous technologies. But we need government now. Fortunately, these same technologies that necessitated government, also provided the means for government to operate.

Somewhat similarly, we haven't needed an AI Nanny so far, because we haven't had sufficiently powerful and destructive technologies. And, these same technologies that apparently necessitate the creation of an AI Nanny, also appear to provide the means of creating it.

The Argument for an AI Nanny

To recap and summarize, the basic argument for trying to build an AI Nanny is founded on the premises that:

1. It's impracticable to halt the exponential advancement of technology (even if one wanted to)
2. As technology advances, it becomes possible for individuals or groups to wreak greater and greater damage using less and less intelligence and resources

3. As technology advances, humans will more and more acutely lack the capability to monitor global technology development and forestall radically dangerous technology-enabled events
4. Creating an AI Nanny is a significantly less difficult technological problem than creating an AI or other technology with a predictably high probability of launching a full-scale positive Singularity
5. Imposing a permanent or very long term constraint on the development of new technologies is undesirable

The fifth and final premise is normative; the others are empirical. None of the empirical premises are certain, but all seem likely to me. The first three premises are strongly implied by recent social and technological trends. The fourth premise seems commonsensical based on current science, mathematics and engineering.

These premises lead to the conclusion that trying to build an AI Nanny is probably a good idea. The actual plausibility of building an AI Nanny is a different matter, of course – I believe it is plausible, but of course, opinions on the plausibility of building any kind of AGI system in the relatively near future vary all over the map.

Complaints and Responses

I have discussed the AI Nanny idea with a variety of people over the last year or so, and have heard an abundance of different complaints about it – but none have struck me as compelling.

> *"It's impossible to build an AI Nanny; the AI R&D is too hard."* – But is it really? It's almost surely impossible to build and install an AI Nanny this year; but as a professional AI researcher, I believe such a thing is well within the realm of possibility. I think we could have one in a couple decades if we really put our collective minds to it. It would involve a host of coordinated research breakthroughs, and a lot of large-scale software and hardware engineering, but nothing implausible according to current science and engineering. We did amazing things in the Manhattan Project because we wanted to win a war – how hard are we willing to try when our overall future is at stake?

It may be worth dissecting this "hard R&D" complaint into two sub-complaints:

- *"AGI is hard":* building an AGI system with slightly greater than human level intelligence is too hard;
- *"Nannifying an AGI is hard":* given a slightly superhuman AGI system, turning into an AI Nanny is too hard.

Obviously both of these are contentious issues.

Regarding the "AGI is hard" complaint, at the AGI-09 artificial intelligence research conference, an expert-assessment survey[72] was done, suggesting that a least a nontrivial plurality of professional AI researchers believes that human-level AGI is

---

[72] http://hplusmagazine.com/2010/02/05/how-long-till-human-level-ai/

possible within the next few decades, and that slightly-superhuman AGI will follow shortly after that.

Regarding the "Nannifying an AGI is hard" complaint, I think its validity depends on the AGI architecture in question. If one is talking about an integrative, cognitive-science-based, explicitly goal-oriented AGI system like, say, OpenCog[73] or MicroPsi[74] or LIDA[75], then this is probably not too much of an issue, as these architectures are fairly flexible and incorporate explicitly articulated goals. If one is talking about, say, an AGI built via closely emulating human brain architecture, in which the designers have relatively weak understanding of the AGI system's representations and dynamics, then the "nannification is hard" problem might be more serious. My own research intuition is that an integrative, cognitive-science-based, explicitly goal-oriented system is likely to be the path via which advanced AGI first arises; this is the path my own work is following.

"It's impossible to build an AI Nanny; the surveillance technology is too hard to implement." – But is it really? Surveillance tech is advancing bloody fast, for all sorts of reasons more prosaic than the potential development of an AI Nanny. Read David Brin's book *The Transparent Society*[76], for a rather compelling argument that before too long, we'll all be able to see everything everyone else is doing.

"Setting up an AI Nanny, in practice, would require a world government." – OK, yes it would ... Sort of. It would require either a proactive assertion of power by some particular party, creating and installing an AI Nanny without asking everybody

---

[73] http://opencog.org/

[74] http://micropsi.com/publications/Bach_AGI_2011.pdf

[75] http://en.wikipedia.org/wiki/LIDA_%28cognitive_architecture%29

[76] http://www.davidbrin.com/transparent.htm

else's permission; or else a degree of cooperation between the world's most powerful governments, beyond what we see today. Either route seems conceivable. Regarding the second cooperative path, it's worth observing that the world is clearly moving in the direction of greater international unity, albeit in fits and starts. Once the profound risks posed by advancing technology become more apparent to the world's leaders, the required sort of international cooperation will probably be a lot easier to come by. Hugo de Garis's most recent book Multis and Monos [77] riffs extensively on the theme of emerging world government.

"Building an AI Nanny is harder than building a self-modifying, self-improving AGI that will retain its Friendly goals even as it self-modifies." – Yes, someone really made this counterargument to me; but as a scientist, mathematician and engineer, I find this wholly implausible. Maintenance of goals under radical self-modification and self-improvement seems to pose some very thorny philosophical and technical problem – and once these are solved (to the extent that they're even solvable) *then* one will have a host of currently-unforeseeable engineering problems to consider. Furthermore there is a huge, almost surely irreducible uncertainty in creating something massively more intelligent than oneself. Whereas creating an AI Nanny is "merely" a very difficult, very large scale science and engineering problem.

"If someone creates a new technology smarter than the AI Nanny, how will the AI Nanny recognize this and be able to nip it in the bud?" – Remember, the hypothesis is that the AI Nanny is significantly smarter than people. Imagine a friendly, highly intelligent person monitoring and supervising the creative projects of a room full of chimps or "intellectually challenged" individuals.

---

[77] http://www.amazon.com/Multis-Multicultured-Monocultured-Towards-Creation/dp/0882801627/

"Why would the AI Nanny want to retain its initially pre-programmed goals, instead of modifying them to suit itself better? – for instance, why wouldn't it simply adopt the goal of becoming an all-powerful dictator and exploiting us for its own ends?" – But why *would* it change its goals? What forces would cause it to become selfish, greedy, etc.? Let's not anthropomorphize. "Power corrupts, and absolute power corrupts absolutely" is a statement about human psychology, not a general law of intelligent systems. Human beings are not architected as rational, goal-oriented systems, even though some of us aspire to be such systems and make some progress toward behaving in this manner. If an AI system is created with an architecture inclining it to pursue certain goals, there's no reason why it would automatically be inclined to modify these goals.

"But how can you specify the AI Nanny's goals precisely? You can't right? And if you specify them imprecisely, how do you know it won't eventually come to interpret them in some way that goes against your original intention? And then if you want to tweak its goals, because you realize you made a mistake, it won't let you, right?" – This is a tough problem, without a perfect solution. But remember, one of its goals is to be open-minded about the possibility that it's misinterpreting its goals. Indeed, one can't rule out the possibility that it will misinterpret this meta-goal and then, in reality, closed-mindedly interpret its other goals in an incorrect way. The AI Nanny would not be a risk-free endeavor, and it would be important to get a feel for its realities before giving it too much power. But again, the question is not whether it's an absolutely safe and positive project – but rather, whether it's better than the alternatives!

"What about Steve Omohundro's 'Basic AI Drives[78]'*? Didn't Omohundro prove that any AI system would seek resources and*

---

[78] http://selfawaresystems.com/2007/11/30/paper-on-the-basic-ai-drives/

power just like human beings?" – Steve's paper is an instant classic, but his arguments are mainly evolutionary. They apply to the case of an AI competing against other roughly equally intelligent and powerful systems for survival. The posited AI Nanny would be smarter and more powerful than any human, and would have, as part of its goal content, the maintenance of this situation for 200 years (200 obviously being a somewhat arbitrary number inserted for convenience of discussion). Unless someone managed to sneak past its defenses and create competitively powerful and smart AI systems, or it encountered alien minds, the premises of Omohundro's arguments don't apply.

"What happens after the 200 years is up?" – I have no effing idea, and that's the whole point. I know what I *want* to happen – I want to create multiple copies of myself, some of which remain about like I am now (but without ever dying), some of which gradually ascend to "godhood" via fusing with uber-powerful AI minds, and the rest of which occupy various intermediate levels of transcension. I want the same to happen for my friends and family, and everyone else who wants it. I want some of my copies to fuse with other minds, and some to remain distinct. I want those who prefer to remain legacy humans, to be able to do so. I want all sorts of things, but that's not the point – the point is that after 200 years of research and development under the protection of the AI Nanny, we would have a lot better idea of what's possible and what isn't than any of us do right now.

"What happens if the 200 years pass and none of the hard problems are solved, and we still don't know how to launch a full-on Singularity in a sufficiently reliably positive way?" – One obvious possibility is to launch the AI Nanny again for a couple hundred more years. Or maybe to launch it again with a different, more sophisticated condition for ceding control (in the case that it, or humans, conceive some such condition during the 200 years).

"What if we figure out how to create a Friendly self-improving massively superhuman AGI only 20 years after the initiation of the AI Nanny – then we'd have to wait another 180 years for the real Singularity to begin!"– That's true of course, but if the AI Nanny is working well, then we're not going to die in the interim, and we'll be having a pretty good time. So what's the big deal? A little patience is a virtue!

"But how can you trust anyone to build the AI Nanny? Won't they secretly put in an override telling the AI Nanny to obey them, but nobody else?" – That's possible, but there would be some good reasons for the AI Nanny developers *not* to do that. For one thing, if others suspected the AI Nanny developers had done this, some of these others would likely capture and torture the developers, in an effort to force them to hand over the secret control password. Developing the AI Nanny via an open, international, democratic community and process would diminish the odds of this sort of problem happening.

"What if, shortly after initiating the AI Nanny, some human sees some fatal flaw in the AI Nanny approach, which we don't see now. Then we'd be unable to undo our mistake." – Oops.

"But it's odious!!" – Yes, it's odious. Government is odious too, but apparently necessary. And as Winston Churchill said, "Democracy is the worst possible form of government, except for all the others." Human life, in many respects, is goddamned odious. Nature is beautiful and cooperative and synergetic – and also red in tooth and claw. Life is wonderful, beautiful and amazing – and tough and full of compromises. Hell, even *physics* is a bit odious – some parts of my brain find the Second Law of Thermodynamics and the Heisenberg Uncertainty Principle damned unsatisfying! I wouldn't have written this chapter in my early 20s, because back then I was more steadfastly oriented toward idealistic solutions – but now, in my mid-40s, I've pretty well come to terms with the universe's persistent refusal to behave in accordance with all my ideals. The AI Nanny scenario is odious in some respects – but human life is odious sometimes,

and we need to be thinking about realistic solutions… It stands to reason that the transcendence of humanity may involve some measure of human odiousness – as well as a rich dose of human beauty!

A Call to Brains

This chapter is not supposed to be a call to arms to create an AI Nanny. As I've said above, the AI Nanny is not an idea that thrills my heart. It irritates me. I love freedom, and I'm also impatient and ambitious – I want the full-on Singularity *yesterday*, goddammit!!!

But I do think it's worth seriously considering whether some form of AI Nanny might well be the best path forward for humanity – the best way for us to ultimately create a Singularity according to our values. At very least, it's worth fleshing out the idea fairly carefully, and weighing it against alternatives.

So this is more of a "call to brains", really. I'd like to get more people thinking about what an AI Nanny might be like, and how we might engineer one. And I'd like to get more people thinking actively and creatively about alternatives.

Perhaps you dislike the AI Nanny idea even more than I do. But even so, consider: Others may feel differently. You may well have an AI Nanny in your future anyway. And even if the notion seems unappealing now, you may well enjoy it quite tremendously when it actually happens.

Oh Brave New World…

22

AGI, Consciousness, Spirituality, Life, the Universe, and Everything

My futurist friend Giulio Prisco has been striving for some time to bring together transhumanism and religion. His Turing Church blog pursues this theme, and he has also organized a series of online conferences.... Conversations with Giulio formed much of my inspiration for writing ***A Cosmist Manifesto*** *– though that book ended up philosophical rather than religious in nature ... in the end I don't quite share Giulio's vision that futurism should, or will, end up having all that much in common with traditional religions. In any case, the following lengthy H+ Magazine article (edited a bit for inclusion here) was also inspired by my ongoing dialogue with Giulio, and represents an attempt by my 2012 self to clarify the various philosophical and spiritual perspectives at play regarding future technologies, the Singularity, and so forth.*

Artificial General Intelligence is a technical topic, but it's also much more than that. It has big implications about humanity's place in the universe – both what humanity's place is logically and scientifically speaking, and what we want it to be normatively. It raises a host of dizzying possibilities regarding our possible futures, including the potential of fusing our minds with AGIs cyborgically, and becoming "more than human" in various senses (some quite strong). It even raises the possibility of understanding the universe and finding answers to the Big Questions of Life, the Universe and Everything – with the help of engineered transhuman minds, to aid us in venturing where the human mind is too weak to travel. And then the possibilities highlighted by folks like Hans Moravec, Hugo de Garis and the SIAI, of human extinction, of our replacement by transhuman AGIs who don't care anymore for humans than humans do for flies or bacteria.

One critical aspect of the relation between AGI and the Big Picture has to do with consciousness. If one thinks AGIs can be conscious in the sense that humans are – or maybe even more broadly and richly conscious – then that certainly informs one's sense of AGI's place in the Cosmos. On the other hand if one things AGIs are intrinsically just feeling-less, experience-less tools, then the story becomes quite different.

There are many, many ways of fitting AGI into one's view of Cosmos and consciousness; a few possibilities are as follows[79]:

- **Physical Monism**. There is a physical universe with a kind of independent, absolute existence (though of a commonsensically peculiar sort, as quantum theory has taught us), and talking of anything besides physical entities and configurations thereof is basically just babbling nonsense.
- **Informational Monism**: The universe basically consists of information, and talking about anything besides information isn't really possible. People are conglomerations of information; and AGIs will ultimately be more complex and capable conglomerations of information. "Soul" and "consciousness" are merely conceptual constructs, i.e. conglomerations of information created by other conglomerations of information to communicate with themselves and each other. "Physical" entities like rocks and electrons are also best thought of as conglomerations of information, since we only know them via the information we perceive about them. In a sense, this means the universe is essentially some kind of computer (though a massively parallel one).

---

[79] Note that the labels in this list are ones I've adopted for my own convenience, rather than necessarily because they're the most standard ones.

- **Quantum Informational Monism**: Like the above, but on quantum theory. The universe basically consists of quantum information, thus is essentially a quantum computer.
- **Cognitive Vitalism**: Human-level AGI is impossible because intelligence requires some kind of "soul" or other special entity, that humans have but engineered entities will never have
- **Experiential Vitalism**: Superhumanly capable AGI is possible, but even so, these AGIs will in a sense be "zombies" without the same kind of consciousness (aka "soul"?) that people have
- **Panpsychism**: Every element of the physical or informational realms is in a sense already "mental", with some sort of spark of awareness; indeed it makes no sense to talk about matter separate from mind. The consciousness associated with a complex system like a brain, body, AGI software system or robot is just a particular manifestation of the universal mental aspect of reality.
- **Spiritual Monism**: Since the material of the universe is spiritual already, as part of its intrinsic nature, each configuration of the universe's matter/information is already associated with its own variety of "soul." Superhuman AGI is possible, and such AGIs will have the same kind of special soul-stuff as humans do – perhaps manifesting it more intensely in some ways, as they develop. Fusing with superhuman AGIs, uploading, or improving our brains in various ways, may be part of our spiritual growth quest.
- **Cosmism**: The universe itself is a massively complex intelligent system, and humans and AGIs are small parts of it, and as systems become more intelligent they have more and more potential to harmonize with the (currently

somewhat mysterious to us) ambient intelligence of the Cosmos

- **The Implicate Order**: There is a sort of "implicate information" structuring the universe, which is different from explicitly observable information – and physical entities have both implicate and explicate aspects. Conscious experience has to do with both aspects, and to the extent there is an "intelligence to the whole universe" it would also have both implicate and explicate aspects. One idea I've been experimenting with is to model the implicate order as a kind of cosmist "questioning process."

Of course this is not intended as a complete list of possible perspectives... And it's also worth noting that these are not mutually exclusive categories, by any means! Some individuals or schools of thought may adhere to more than one of these perspectives.

Sometimes I hear people express the naïve thought that AGI research is naturally aligned with a particular philosophical perspective, usually the views I refer to above as physical or informational monism. But I don't think any such natural alignment exists. I've observed AGI researchers and AGI-focused transhumanist thinkers to take a great variety of perspectives.

In this chapter I'll briefly run through these views I've listed, with more emphasis on breadth than depth. I don't want to turn this into a book on cosmic philosophy! – But I do want to indicate the wonderful diversity of ways in which the notion of AGI is already being interwoven with various human thought and belief systems. One can expect the weaving of AGI into these various human conceptual networks to get even richer once advanced AGI is actually created and interacts with us on a regular basis!

I'll spend relatively little time on those perspectives that are already extremely well-known due to their commonality, and dwell more on those views I find most interesting! Also, note that two of my own favorite perspectives appear at the end of the chapter (Cosmism and the Implicate Order... My other favorites of the bunch being Informational Monism and Panpsychism). So if you don't read to the end you'll miss the good stuff!

Physical Monism

What I refer to as "physical monism" is the view that the physical world is in some sense absolutely real and out there and existent – and everything else is a bunch of malarkey. Intelligence, mind and consciousness and experience are viewed as patterns or configurations (or whatever) of physical entities like particles or waves.

To my mind, this is a potentially useful (though rather limited) perspective for getting through everyday life, but somewhat deficient as a real intellectual theory.

After all, how do we know about this supposedly "absolutely existent" physical world? Either we take it on faith, like a prototypical religious person's belief in God... Or we infer it from various observations, i.e. various pieces of information. But if it's the latter – then isn't the more fundamental reality associated with the *information*, rather than with the physical world whose "existence" we infer from the information?

A classic expression of this view is G.E. Moore's observation that when you kick a rock, you *know* it's real, and philosophical babbling becomes irrelevant in the light of this direct evidence of the rock's reality. But what is it that you actually know is real, when you kick the rock? The definite experienced reality is attached to the feelings coming to you from your foot (i.e. feelings that you have learned to attach to the concept/percept network called "foot"), and the sensations coming into your eye when you look at your rock and the foot. Actually there might be no rock present in physical reality at all – your brain might be connected to an apparatus that makes you *think* you're kicking a rock.

And this leads up to the next perspective in my list, which I find deeper and more interesting...

Informational Monism

The view of the world as bottoming out in some absolute physical reality seems hopelessly naïve to me -- but the view of the world as bottoming out in *information* seems much less so. I don't *quite* adhere to this perspective myself, but nor do I know of any rational, scientific arguments against it. AGI researcher Joscha Bach, whom I interviewed above, has described the perspective well:

> *We grow up with the illusion of a direct access to an outside world, and this intuition is reflected in the correspondence theory of truth: our concepts derive their meaning from their correspondence to facts in a given reality. But how does this correspondence work? According to our current understanding of biology, all access to the world is mediated through a transfer of data, of bits (i.e., discernible differences) expressed by electrical impulses through sensory and motor nerves. The structure of the world, with percepts, concepts, relations and so on, is not part of this data transfer, but is constructed by our minds. It represents encodings over the regularities found in the data patterns at the mind's interface. In theory, all the data entering my mind over my lifetime could be recorded as a finite (but very long) vector of bits, and everything that I consider to be knowledge of the world is a more concise re-encoding of parts of this vector.*
>
> *Of course, even the concept that the world is mediated through sensory nerves is not directly accessible. It is an encoding, too (it just happens to be the best encoding that we found so far). And because we cannot know the "real"*

structure behind the data patterns at our mind's systemic interface, the correspondence theory of truth does not work, outside the abstract and pure realm of mathematics.

But what about information itself: how can we know about bits, data and encodings? Fortunately, these are mathematical concepts. They can be defined and operated upon outside of any empirical world. For instance, the theory of Natural Numbers does not need anything material (like piles of apples and oranges) to work: It follows automatically from the completely abstract and theoretical set of Peano's axioms. But Natural Numbers can be used to encode some aspect of apples and oranges--their cardinality, and their behavior with respect to addition and subtraction.

The space of mathematics is self-contained, but to explore it, we need a certain kind of information processing system, which is a mathematical entity too. Minds are part of the class of information processing systems that can perform mathematics (at least to some extend), but minds can do even more: they can conceptualize a world, reflect, plan, imagine, anticipate, decide, dream, interpret themselves as persons, be in emotional states, attach relevance to concepts and so on. The concept of mind is the one we attach to ourselves, we use it to encode that part of the information vector that we consider to be us.

With respect to our understanding of it, the abstract theory of minds is still in its infancy. Our common-sense understanding of what it takes to be a mind is good enough to use it as an encoding concept (i.e., to use it as a conceptual framework that allows the interpretation of parts of the world, as people, self, mental states, emotions, motives, beliefs and so on). Yet, our concept of minds is incomplete, muddled and likely inconsistent. It is possible, however, taken enough time, effort and brain power, to define all the aforementioned capabilities of minds down to the nitty-gritty detail that would make it the mathematical equivalent of a theory of Natural Numbers: a formal and complete theory of what it takes to be a mind. We can express this theory, for instance, as a computer program. And this project is exactly what Artificial Intelligence is about.

I think this makes far more sense than physical realism, or mind/matter dualism. It has led some thinkers to posit that the universe is basically a giant computer. While I don't quite embrace this myself, I do think the model of the universe as a giant computer has a lot to teach us.

Joscha and I have argued about this extensively face-to-face, and as I told him in our conversations, I agree with him that **insofar as it's investigable using science, the universe may be considered as consisting solely of information**. The difference between our perspectives is that I don't think all aspects of the universe are scientifically investigable.

Science, by its nature, is about gathering finite sets of finite-precision observations – i.e. the whole corpus of scientific knowledge consists of some finite set of bits. But there's no reason to believe that the whole universe consists of some finite set of bits. We may not be able to measure anything else via the means science, but this just means that only information is "scientifically existent", not that "scientific existence" is the only kind of existence!

When we last discussed the matter, he was willing to concede this logical point, but also said (to paraphrase more or less loosely) he felt there was no point in attempting to talk or think about forms of existence beyond the scientifically measurable realm. And this difference of perspective between us doesn't affect our ability to collaborate on AGI work at all, since our AGI engineering efforts do in fact concern the explicit manipulation of scientifically measurable entities!

Quantum Informational Monism

A variant on the above perspective holds that the universe basically consists of quantum information. This seems to be, for example, the view of Seth Lloyd, quantum computing pioneer and author of the fabulous book *Programming the Universe*. In the same sense that informational monism deems the world a classical computer, this perspective deems the world a quantum computer.

In the quantum informational monist view, classical information is an approximation to quantum information, relevant in situations of minimal quantum coherence. From the classical informational monist view, on the other hand, quantum theory itself is just a computational explanation of some classical information that's been gathered from measuring instruments (which are themselves known via information gathered via other measuring instruments, either artificial ones or the biological ones we call our senses). I.e., quantum theory is a classical computer program for getting from some classical information to some other classical information.

Joscha Bach has summarized the informational monist view of quantum computation as follows:

> *Everything that enters the human mind can be reduced to a finite number of discernible differences. In this sense, the most primitive encoding of the universe, as presented to an individual observer, would be an astronomically large, finite vector of bits. Obviously, our minds can do better than that: much of the structure found in that vector conforms to a three-dimensional space, filled with objects. These objects may influence other objects in their vicinity, by touching them, radiating upon them and so on. Thus, we arrive at classical physics as a way of encoding the universe. Classical physics, when expressed rigorously, is a self-contained mathematical theory; objects in classical physics are entities with properties and modes of interaction that are*

reducible to calculations. As it turns out, classical physics can not only be reduced to computation, it is in turn powerful enough to explain how computation is possible (i.e., information processing in general): using classical physics, we can conceive of a computer that can simulate classical physics.

When we look upon our basic universe input vector more closely, however, we find that the model of classical physics is flawed: it is not totally consistent with the available data. What we take to be objects does not only interact locally, but occasionally over distances in space and time, and what we take to be a defined state of a microscopic object can only be described as a space of possible states, which it occupies all at once. If we care about this discrepancy, we have to abandon the notion of the classical universe, and adopt a theory that accommodates the observations, i.e., quantum mechanics.

Even the quantum mechanical universe is just our way of encoding the finite string of bits we began with, so of course it is computational. But quantum computation allows to accommodate computers that perform some kinds of computation vastly more efficient, which means that a computer designed along classical principles might be too slow to practically simulate a complex quantum mechanical system.

At this point, it is not entirely clear if a computational theory of the mind would have to be formulated along the lines of quantum computation, or if classical computation is sufficient. But there is practically no evidence that the information processing performed by the neurons in our brains would somehow crucially depend on non-local or quantum-superpositional effects, or that people can somehow perform computations that would require quantum computing. Thus, even though classical computers are too limited for detailed low-level simulations of our universe, they are likely probably perfectly adequate for a detailed low-level simulation of a mind.

For Joscha, in other words, it's bits that are primary. Classical physics is one way of explaining bits, and quantum physics is another way of explaining bits. Either classical or quantum physics is ultimately expressible as a mathematical construct, i.e. as a set of formulas for manipulating symbols, used to predict future bits from past bits. So in Joscha's informational monist view, the ultimately reality is the bits and not the classical or quantum physical explanation of bit patterns. On the other hand, from a quantum computationalist view, if the assumption of the quantum model of reality is the simplest way to explain the stream of observed bits, then this merits the assignation of some fundamental reality to the quantum constructs involved in the quantum model of reality.

Experiential Vitalism

The perspectives I've considered so far in this chapter are basically "scientific" in nature, focused on things like physics and information theory. But there are also some serious AGI researchers who take quite different views, drawing more from the religious or spiritual side of life.

I'm not a religious person by nature or culture – according to facile high-level classifications of religious belief, I've veered between "atheist", "agnostic" and "spiritual but not religious" in various forms at various times of my life. However, I've always been fascinated by religious peoples' beliefs and attitudes, feeling that they do address important aspects of the universe that science (so far, at least) has given us little insight about.

After all, no matter how much you value math and science, they can't tell you everything. Mathematical derivations start with assumed axioms; and as David Hume was the first to carefully argue) science requires some inductive bias or it can't infer anything specific from observed data. Everybody's got to assume something to get by in the world, whether they realize it or not. So it's interesting to look at the great variety of things that different people and cultures assume.

Even those who, like Zen masters or Sufis, like to talk about assuming nothing – even these folks still implicitly act as if they're assuming something, as they go about their lives. When the Zen master opens his refrigerator to get some water, he is implicitly assuming there will be water in there – instead of for example, a lion, in which case he might bring his gun with him to the fridge rather than a cup. And the various implicit assumptions he makes throughout his life network together and evolve in complex ways, just like with everybody else. Both the implicit and the explicit assumptions people make are quite interesting.

However, I do find it interesting to understand what truly religious people think about AGI and other transhumanist topics. In an attempt to fulfill my curiosity on this, last year I did some H+ Magazine interviews with a few religious and spiritual types about AGI, transhumanism and religion. I won't include the whole interviews here because that would bring us too far afield; but I'll give some brief highlights that may give you a sense for how some others are thinking about this…

For starters: Changle Zhou, the dean of the Cognitive Science Department at Xiamen University – where I'm an adjunct research professor, and where a team of students are working on some OpenCog-related AI software projects – is also an experienced Zen Buddhist practitioner. He reports with a grin that he stopped his formal Zen study some years back, when his Zen master declared him Enlightened! He's also the author of a book, in Chinese, about the relation between Zen and Science. It was both a privilege and a source of considerable entertainment for me to interview him about his views on the relation between AI and Zen…

Consciousness, as I understand it, has three properties: self-referential, coherent and qualia. Even if a robot becomes a zombie that acts just like it possesses consciousness, it won't really possess consciousness – because the three properties of consciousness cannot all be realized by the reductive, analytic approach, whereas AI is based on this sort of approach.

...

In Zen we say that SUCHNESS is the nature of everything. For a conscious human being, its Suchness is its consciousness; but for material things, their Suchness is their properties such as shape, substance or weight but not consciousness.

...

The Chinese word for "suchness" is "真如" or "自性", i.e. "buddhahood", also known as "the foundational 识" or "种子识" or "Alayavijnana" etc.... It is the name for the nature of everything.

...

The awareness (enlightenment) of Zen is beyond all concepts, but all the approaches we use for building robots and AI systems, and all the behaviors of robots and AI systems, are based on concepts

Perhaps these brief excerpts give you the essence of our dialogue, and our disagreement. He believes that super-capable robots and other amazing futurist technologies are likely to exist – but he sees any robots that human build, as essentially extensions of human intelligence. He sees "natural" intelligences like humans, as possessing a certain special quality – Suchness – that mechanical objects like robots will never possess.

I call this sort of perspective "experiential vitalism." That is: the belief that, while the practical functions of intelligence can be carried out by digital computers or other engineered systems, there's some essential crux of consciousness that human mind/brains possess that these other systems never will.

Cognitive Vitalism

A related perspective on AGI and the cosmos, fairly common in the general population though not so much among scientists, is that human brains contain some kind of special quality, which lies outside the domain of empirical science and is responsible for some key aspects of human intelligence.

Perhaps the most articulate advocates of this perspective that I've encountered id Selmer Bringsjord, an AI researcher and logician who is also a devout Christian. Where Changle says Suchness, Selmer says Soul – but their perspectives seem closely related. However, there's a key difference – because Selmer also argues that some *functional* aspects of human-level intelligence are reliant on the Soul, which digital computers will never possess. This, he believes, is the core reason why the AI field has not yet led to software with human-level general intelligence.

Selmer goes on to associate the soul with non-Turing hypercomputation, a kind of computing that cannot be achieved via physical computers built according to any kind of currently known physics, and that also cannot be measured using empirical science as currently understood. That is: Science as we know it consists of the collection of "scientific data" which consists of finite sets of bits (two other ways to phrase "finite sets of bits" would be "finite amounts of information" or "finite sets of finite-precision numbers"), and then the extrapolation of these to predict the outcomes of future experiments, which will also take the form of finite sets of bits. There is no scientific experiment, conductable within the contemporary understanding or practice of science, that would be able to identify a hypercomputable process, or distinguish it from a conventional computing process. In that sense, hypercomputing is a non-empirical concept – beyond the realm of measurement!

And here the philosophy gets tricky.

If hypercomputers are not measurable, in what sense can we know about them at all? What sense does it make to talk about them? Well, they can be described using mathematics!

But then, what is this mathematics? In practice it's just the marking-down of mathematical symbols on paper or computer screens – i.e. it's a game played by humans with finite sets of bits.

But in the Platonist philosophy of mathematics, mathematical entities are envisioned to have some fundamental reality going beyond the notations that we make to indicate them. So if one accepts a Platonist view that math constructs have their own reality beyond empirical science or human communication, then one may say that hypercomputable entities exist in this abstract Platonic math-space, and that Soul also exists in (or at least is better represented in terms of, compared to any representation in terms of empirical science) this abstract Platonic math-space. And digital computers that we build, which manipulate finite bit sets based on scientific theories inferred from scientific data that's comprised of finite bit sets, live in a much smaller and more impoverished region of abstract Platonic math-space, not touching the region needed to talk meaningfully about human Soul, or human-level intelligence.

One of the more frustrating, and gutsier, things about this perspective is that it ultimately places the understanding of human-level general intelligence outside the domain of science – though not necessarily outside the domain of mathematics. It suggests that we can intuitively apprehend human-level general intelligence using our trans-scientific hypercomputation capabilities, but can never test these apprehensions scientifically. It also raises the possibility that maybe we could somehow build an AGI using intuition rather than science. I.e., if the world has hypercomputable aspects, and our minds do also, then maybe the hypercomputable aspects of our minds could intuitively tell us how to shape the hypercomputable aspects of the world, and we could make a physical AGI system in a manner not determined by any finite bit sets of measurements.

But so far as I know Selmer is not trying to pursue any kind of recondite Soul-guided AGI engineering of this nature. Rather, his concrete AGI work is focused on the implementation of logic systems in digital computers – and on seeing exactly how far one can push this sort of methodology before it runs into the fundamental limits that he believes to exist.

I don't know of any simple non-technical write-up of these interesting notions, but if you have a bit of math and computing background, you may enjoy Selmer's book *Superminds: People Harness Hypercomputation, and More*.

Panpsychism

Panpsychism occurs in various forms, but in the broad sense it refers simply to the idea that mind is a fundamental feature of the universe and each of its parts, rather than something that is the exclusive property of specific kinds of systems like humans, other higher animals, intelligent computer programs, etc.

Though not a common view in contemporary Western society, philosophy or science, panpsychism does have a long history in historical Western philosophy, encompassing thinkers like Leibniz, James, Whitehead, Russell, Fechner and Spinoza. A host of recent books treat the topic, including Skrbina's *Mind that Abides: Panpsychism in the New Millienium* and Strawson's *Consciousness and its Place in Nature*.

Panpsychism also has a long and rich history in Eastern philosophy, e.g. the modern Vedantic thinker Swami Krishnanda observes

The Vedanta philosophy concludes that matter also is a phase of consciousness and objects of knowledge embody in themselves a hidden potential of consciousness which is also the Self of the perceiving subject, enabling experience in the subject. The subject-consciousness (Vishayi-chaitanya) is in a larger dimension of its own being as universality and all-pervadingness beholds itself in the object-consciousness (Vishaya-chaitanya), thereby reducing all possible experience to a degree of universal consciousness. Experience is neither purely subjective nor entirely objective; experience is caused by the universal element inherent in both the subject and the object, linking the two terms of the relation together and yet transcending both the subject and the object because of its universality.

Advocates of panpsychism point out that alternative theories of mind and consciousness are riddled with problems and inconsistencies, whereas panpsychism is simple and coherent, its only "problem" being that it disagrees with the intuition of many modern Western folk. Most current theories of consciousness involve mind and awareness somehow emerging out of non-sentient matter, which is conceptually problematic. Philosopher Galen Strawson has recently lamented the basic senselessness of the notion that mental experience can emerge from a wholly non-mental, non-experiential substrate: "I think it is very, very hard to understand what it is supposed to involve. I think that it is incoherent, in fact..."

Dualist theories in which the mind-realm and the matter-realm are separate but communicating also run into difficulties, e.g. the problem that (put crudely) the mind-realm must be utterly undetectable via science or else in effect it becomes part of the matter-realm. Panpsychism holds that everything in the world has mental extent, similar to how it has spatial and temporal extent, which is a simple proposal that doesn't give rise to any conceptual contradictions.

Some have objected to panpsychism due to the apparent lack of evidence that the fundamental entities of the physical world possess any mentalistic properties. However, this lack of evidence may easily be attributed to our poor observational skills. By analogy, humans cannot directly detect the gravitational properties of small objects, but this doesn't render such properties nonexistent. And in appropriate states of consciousness, humans *can* directly apprehend the consciousness of objects like rocks, chairs or particles, a fact driven home forcefully by Aldous Huxley in his classic book *The Perennial Philosophy*.

Panpsychism is not without its difficulties, e.g. the "combination problem," first raised by William James – which in essence wonders: if everything is conscious, how does the consciousness of a whole relate to the consciousnesses of its parts? How does the brain's consciousness come out of the consciousnessess of its component neurons, for example?

But this doesn't seem a problem on the order of "how does consciousness emerge from non-conscious matter", it seems more a technical issue. A large variety of qualitatively different part-whole relationships may exists, as physicists have noted in the last century. Quantum mechanics has made clear that systems are not simply the sum of their parts but can sometimes exhibit properties that go beyond those of the parts and which cannot be detected by examining the parts in isolation. And black hole physics has shown us the possibility of wholes (black holes) that totally lose most of the properties possessed by their parts and render the parts in accessible (a black hole has only the properties of mass, charge and spin, regardless of the other properties possessed by the objects that combined to form the black hole). The nature of part-whole relationships in panpsychism certainly bears further study, but merely appears subtle, not incoherent. And the emergent, holistic aspect of consciousness is well known in Eastern thought, e.g. Swami Krishananda says that:

> *The three states of waking, dream and sleep, through which we pass in our daily experience, differ from one another, and yet a single consciousness connects them, enabling the individual to experience an identity even in the otherwise*

differentiatedness of these states. Since consciousness links the three states into a singleness of experience, it is immanent in them and yet transcends them, not capable of identity with any of them.

In short, the panpsychist view of consciousness has a long history in both Eastern and Western philosophy, and has no glaring conceptual problems associated with it, the main difficulty with it being that most people in contemporary Western cultures find it counterintuitive. At least one of the authors has found it a useful guide for thinking about the mind, perhaps largely because it doesn't contain any confusing inconsistencies or incoherencies that "get in the way" of analyzing other issues involved with machine consciousness, such as reflective consciousness, self and will.

Spiritual Monism

Panpsychism holds that everything in the cosmos has at least a spark of mind, of cosciousness, in it. Quite often – though not always – this perspective comes along with a more religious view, which holds that everything in the cosmos has a spark of *God* in it (in some sense or another). So that ultimately, everything is part of the Mind of God. I call this "spiritual monism." Of course this may be interpreted in many, many different ways and it would be out of place to review them all here, but I can't resist giving one particularly fascinating example.

Around the same time I interviewed Changle, I also did a long and fascinating interview with Lincoln Cannon, the leader of the Mormon Transhumanist Association. As well as advocating a sort of Mormon spiritual monism, he advocats the intriguing position that Mormonism is the "most transhumanist religion" – because it explicitly advocates human beings improving themselves incrementally until they effectively become gods. It also contains the idea that God used to be an ordinary being like us, until he self-improved and became, well, transhuman...

What I find fascinating when talking to Lincoln, is how thoroughly he's integrated transhumanism – AGI, nanotechnology, mind uploading and all the rest – into his Mormon world view. It's not as though he keeps Mormonism and transhumanism in different parts of his brain, one personal and one intellectual or whatever – rather, for him they're all parts of one big conceptual complex! And in this conceptual complex, unlike Changle's, there is plenty of room for computer programs and robots that have the same kind of intelligence, consciousness and spirituality as humans:

Ben

Oh, and one more thing I just can't resist asking... In your perspective could an intelligent computer program have a soul? Could it have consciousness? Could an intelligent computer program become a God, in the same sense that a person could? Will AIs be able to participate in the collective deification process on the same status as humans?

Lincoln

In Mormonism, "soul" is used to describe the combination of spirit and body, rather than just spirit. ... I think computer programs already have spirits, or actually ARE spirits.

In Mormon cosmology, God creates everything spiritually before physically, organizing and reorganizing uncreated spirit and matter toward greater joy and glory. All things have spirits. Humans have spirits. Non-human animals and even the Earth have spirits, and will be glorified according to the measure of their creation, along with us. Many Mormons also anticipate that the day will come when, emulating God, we learn to create our own spirit children. Spirit is in and through all things. Recall, too, that Mormons are philosophical materialists (not dualists), so even spirit is matter, which God organizes as the spiritual creation of all things. So far as I'm concerned, spirit as described by Mormonism is information, and software engineering is spiritual creation. We are already engaged in the early stages of the creation of our spirit children. Taking a step back, consider how this adds perspective to the problem of evil: what justifies our development of artificial intelligence in an evil world?

Cosmism

Another take on the relation of AGI to the cosmos is contained in the broad philosophy of "Cosmism" – a term originated by Konstantin Tsiolokovsky and other Russian Cosmists in the late 1800s, and then borrowed by myself and Giulio Prisco in 2010 to denote a closely related futurist philosophy, more tailored for the modern era. Rather than positing a fundamental theory about the stuff of the universe or the makeup of mind, Cosmism posits an attitude toward life, technology and the world, which includes an attitude toward AGI and the cosmos and refinement of our understanding of their nature and relationship.

I wrote a short book in 2010 called *A Cosmist Manifesto,* presenting my views on Life, the Universe and Everything, and drawing together thoughts about AGI and other advanced technologies with my take on Zen Buddhism and other spiritual philosophies. As I state at the start of the book,

> *By Cosmism I mean: a practical philosophy focused on enthusiastically and thoroughly exploring, understanding and enjoying the cosmos, in its inner, outer and social aspects*
>
> *Cosmism advocates*
>
> - *pursuing joy, growth and freedom for oneself and all beings*
> - *ongoingly, actively seeking to better understand the universe in its multiple aspects, from a variety of perspectives*
> - *taking nothing as axiomatic and accepting all ideas, beliefs and habits as open to revision based on thought, dialogue and experience*

Near the beginning of the Cosmist Manifesto is the following list of high-level principles, which was initially written by Giulio Prisco and then edited by myself:

Ten Cosmist Convictions

1. Humans will merge with technology, to a rapidly increasing extent. This is a new phase of the evolution of our species, just picking up speed about now. The divide between natural and artificial will blur, then disappear. Some of us will continue to be humans, but with a radically expanded and always growing range of available options, and radically increased diversity and complexity. Others will grow into new forms of intelligence far beyond the human domain.

2. We will develop sentient AI and mind uploading technology. Mind uploading technology will permit an indefinite lifespan to those who choose to leave biology behind and upload. Some uploaded humans will choose to merge with each other and with AIs. This will require reformulations of current notions of self, but we will be able to cope.

3. We will spread to the stars and roam the universe. We will meet and merge with other species out there. We may roam to other dimensions of existence as well, beyond the ones of which we're currently aware.

4. We will develop interoperable synthetic realities (virtual worlds) able to support sentience. Some uploads will choose to live in virtual worlds. The divide between physical and synthetic realities will blur, then disappear.

5. We will develop spacetime engineering and scientific "future magic" much beyond our current understanding and imagination.

6. Spacetime engineering and future magic will permit achieving, by scientific means, most of the promises of religions — and many amazing things that no human religion ever dreamed. Eventually we will be able to resurrect the dead by "copying them to the future".
7. Intelligent life will become the main factor in the evolution of the cosmos, and steer it toward an intended path.
8. Radical technological advances will reduce material scarcity drastically, so that abundances of wealth, growth and experience will be available to all minds who so desire. New systems of self-regulation will emerge to mitigate the possibility of mind-creation running amok and exhausting the ample resources of the cosmos.
9. New ethical systems will emerge, based on principles including the spread of joy, growth and freedom through the universe, as well as new principles we cannot yet imagine
10. All these changes will fundamentally improve the subjective and social experience of humans and our creations and successors, leading to states of individual and shared awareness possessing depth, breadth and wonder far beyond that accessible to "legacy humans"

This is followed by a longer list of a few dozen principles and hypotheses, which are then elaborated in more detail – but, well, if you want to know all that, just read the *Cosmist Manifesto* book!

In the Cosmist, view AGI is just one (important) part of the ongoing process of intelligence spreading itself through the Cosmos – a process that we and our intelligent software creations are products of, agents of, and passengers upon.

In late 2010 Giulio asked me to give a talk at an online "Turing Church" workshop he organized, on the "Cosmist Manifesto" theme… I spoke off the cuff as usual, but beforehand I wrote down some notes so I'd have them there to look at, just in case I drew a blank during the actual workshop (which didn't happen, thankfully!). Here are the notes I made, which are somewhat similar to what I actually said:

> *The relation between transhumanism and spirituality is a big topic, which I've thought about a lot -- right now I'll just make a few short comments. Sorry that I won't be able to stick around for this whole meeting today, I have some family stuff I need to do, but I'm happy to be able to participate at least briefly by saying a few remarks.*
>
> *Earlier this year I wrote a book touching on some of these comments, called "A Cosmist Manifesto" -- I'm not going to reiterate all that material now, just touch on a few key points.*
>
> *The individual human mind has a tendency to tie itself in what the psychologist Stanislaw Grof calls "knots" – intricate webs of self-contradiction and fear, that cause emotional pain and cognitive confusion and serve as traps for mental energy. Ultimately these knots are largely rooted in the human self's fear of losing itself – the self's fear of realizing that it lacks fundamental reality, and is basically a construct whose main goals are to keep the body going and reproducing and to preserve itself. These are some complicated words for describing something pretty basic, but I guess we all know what I'm talking about.*
>
> *And then there are the social knots, going beyond the individual ones… The knots we tie each other up in…*
>
> *These knots are serious problems for all of us – and they're an even more serious problem when you think about the potential consequences of advanced technology in the next decade. We're on the verge of creating superhuman AI and molecular nanotech and brain-computer interfacing and so forth – but we're still pretty much fucked up with*

psychological and social confusions! As Freud pointed out in Civilization and its Discontents, we're largely operating with motivational systems evolved for being hunter-gatherers in the African savannah, but the world we're creating for ourselves is dramatically different from that.

Human society has come up with a bunch of different ways to get past these knots.

One of them is religion – which opens a doorway to transpersonal experience, going beyond self and society, opening things up to a broader domain of perceiving, being, understanding and acting. If you're not familiar with more philosophical side of the traditional religions you should look at Aldous Huxley's classic book "The Perennial Philosophy" – it was really an eye-opener for me.

Another method for getting past the knots is science. By focusing on empirical data, collectively perceived and understood, science lets us go beyond our preconceptions and emotions and biases and ideas. Science, with its focus on data and collective rational understanding, provides a powerful engine for growth of understanding. There's a saying that "science advances one funeral at a time" – i.e. old scientific ideas only die when their proponents die. But the remarkable thing is, this isn't entirely true. Science has an amazing capability to push people to give up their closely held ideas, when these ideas don't mesh well with the evidence.

What I see in the transhumanism-meets-spirituality connection is the possibility of somehow bringing together these two great ways of getting beyond the knots. If science and spirituality can come together somehow, we may have a much more powerful way of getting past the individual and social knots that bind us. If we could somehow combine the rigorous data focus of science with the personal and collective mind-purification of spiritual traditions, then we'd have something pretty new and pretty interesting – and maybe something that could help us grapple with the complex issues modern technology is going to bring us in the next few decades

One specific area of science that seems very relevant to these considerations is consciousness studies. Science is having a hard time grappling with consciousness, though it's discovering a lot about neural and cognitive correlates of consciousness. Spiritual traditions have discovered a lot about consciousness, though a lot of this knowledge is expressed in language that's hard for modern people to deal with. I wonder if some kind of science plus spirituality hybrid could provide a new way for groups of people to understand consciousness, combining scientific data and spiritual understanding.

One idea I mentioned in the Cosmist Manifesto book is some sort of "Confederation of Cosmists", and Giulio asked me to say a little bit about that here. The core idea is obvious – some kind of social group of individuals interested in both advanced technology and its implications, and personal growth and mind-expansion. The specific manifestation of the idea isn't too clear. But I wonder if one useful approach might be to focus on the cross-disciplinary understanding of consciousness – using science and spirituality, and also advanced technologies like neuroscience and BCI and AGI. My thinking is that consciousness studies is one concrete area that truly seems to demand some kind of fusion of scientific and spiritual ideas… So maybe focusing on that in a truly broad, cross-tradition, Cosmist way could help us come together more and over help us work together to overcome our various personal and collective knots, and build a better future, and all that good stuff…

Anyway there are just some preliminary thoughts, these are things I'm thinking about a lot these days, and I look forward to sharing my ideas more with you as my thoughts develop

Cosmism doesn't ask us to commit to viewing the universe as information or quantum information or hypercomputation or God-stuff or whatever – it asks for a broader sort of commitment, to an attitude of joy, growth, choice and open-mindedness. Adoption of this attitude may lead to a variety of different perspectives on AGI as it and related technologies evolve. Science in its current form, and religion and philosophy in their current forms, may turn out to be overly limited for the task of understanding (human or artificial) mind; if so, by actively engaging with the world and studying and engineering things, and by reflecting on ourselves carefully and intelligently, we will likely be able to discover the next stage in the evolution of collective thinking...

The Implicate Order

Finally, at risk of leaving you thinking I'm totally nuts, I'm going to share with you some more recent thinking I've been doing, going beyond the ideas in the *Cosmist Manifesto* in a sense. These are half-baked ideas at this stage -- but who knows, maybe some reader will encounter them and have and publish an idea that will help my own thinking along.

The great quantum physicist David Bohm, when he turned more philosophical in his later years, posited the notion of the "implicate order" – i.e. an aspect of the universe that implicitly underlies all things, but isn't in itself scientifically measurable or sensorially perceptible. The explicate order that we can see and measure, in some sense emerges from the implicate order (and then folds back into it, contributing to it).

There are relations between this notion and panpsychism, in that the "spark of mind" implicit in something may be equated to (or at very least associated with) the "implicate aspect" of that thing. Bohm also connected the implicate order with quantum mechanics, though in a manner that I never fully understood from his writings. Sometimes it seemed he wanted us to look at quantum logic as a sort of interface between the implicate and explicate orders. Not that the explicate order uses classical logic and the implicate order uses quantum logic; but rather that quantum logic captures some aspects of the explicate/implicate interaction that classical logic misses.

Recently I've begun thinking about the implicate order from a different perspective, and looking at models of the implicate order as a "logic of questions" rather than a logic of answers. I'm experimenting with modeling the implicate order as something I call QP, a questioning process – not a process of questioning anything in the everyday world, but rather a process of questioning itself. To quote a manuscript I wrote on this a while ago (tentatively titled "?"),

> *If I had to summarize QP in a brief phrase of (almost) ordinary English, I suppose I'd go with something like "the process of a complex, autopoietic pattern/process-system growing and developing via self-referentially (and joyfully, autonomously and interconnectedly) self-questioning." Sorry if that sounds like gobbledygook! It makes a lot of sense to me, and I hope it will make a lot of sense to you after you finish the book!*

QP... The process of questioning. Questioning everything, including the process of questioning everything – and so forth! What I've been studying is how one might model the universe as something that fundamentally emerges from this kind of self-questioning process.

Another famous quantum physicist, John Wheeler, speculated about the possibility of deriving quantum mechanics and general relativity theory (two great physics theories that remain un-unified, leaving modern physics in a state of unacceptable contradiction) from some sort of statistical analysis of the space of logical propositions. So that physics would emerge from a "pre-geometry" made of logic. My QP approach is actually somewhat similar, except that I'm looking at a logic of questions rather than a logic of answers... And thinking about the emergence of the mental as well as physical universes.

Science, after all, is a questioning process – it's about questioning the universe, and about questioning science itself. Every scientific theory brings new ideas and tools used to question its own self, which is why every scientific theory ultimately leads to its own destruction/transcension. The scientific method itself is not a constant, it's constantly being revised, because it invariably leads to relentless questioning which undermines its own foundations.

And while institutionalized religions may seem to have more to do with obeying than with questioning, Gnostic Christianity was precisely about trying to question and understand everything for oneself, to know "God" directly. Jainist Buddhism and Sufism shared the same aspect – Jainists were trained to greet their every thought, belief or concern with the response "Not this! Not this!"

In the QP perspective I elaborate in "?", the questioning processes of science and gnostic religion are examples and instantiations of the broader questioning process woven into the fabric of existence.

As I write this, I'm reminded of my Hong Kong friend Gino Yu, who likes to throw up his hands and grin and ask his friends, gesturing at the world around, and at himself and the rest of us: "What is this??" When I mentioned QP to Gino he simply said "Oh, that's just the Socratic method."

What is this, indeed??

AGI will not tell us "what is this", in any definitive way. AGI will get rid of some of our old questions, and replace them with new questions!

The Ten Cosmist Convictions describe the future of humanity – the future of the ongoing growth and unfolding of the universe, the ramification of the Cosmos into configurations as far beyond current human life as we are beyond the proto-life in the oceans of early Earth, or the lifeless molecules spinning through the void of space prior to the formation of planets. The QP train of thought, tries to dig deeper into this growth and unfolding process – looking at the ongoing unfolding and growth of the universe as a process of relentless self-questioning. It looks at the universe as a big mind, which is constantly asking itself "What is this? What am I doing?" and in doing so is changing itself, bringing about new forms like planets and proto-life and humans and AGI systems.

But please be assured: You don't have to follow me into these peculiar thought-domains to appreciate my AGI work, or my projections about the future of technology! Any more than you have to agree with my panpsychist view of consciousness to think a completed OpenCog system will be conscious – you may have your own conception of consciousness and your own way of applying it to OpenCog.

One thing I'm quite sure of is: None of us humans really knows what's going on!

Conclusion

So, at the end of this romp through strange ideas and perspectives, what's the take-home point about AGI and the Cosmos?

As you already know, I'm not a Mormon – I'm not religious at all… I'm not even really a Jew, in terms of beliefs, in spite of my Jewish racial background. And unlike my colleague Selmer, I don't place much stock in the hypercomputable Soul.

I feel like existing religious and spiritual notions of "God" and "Soul" are getting at important aspects of the universe which science misses (and probably will always miss, due to its foundation in data-sets comprising finite sets of bits); yet I also feel like they're tangled up with a lot of superstitious beliefs and historical "cruft", which makes me want to consider them more as general inspiration than as foundations for my thought and understanding.

I don't know whether classical or quantum information is ultimately a better model of the universe; I think there are a lot of unknowns thereabouts, which will be resolved as science, math and philosophy unfold.

I strongly gravitate toward some form of panpsychism, but I'm not exactly sure what kind. When consciousness theorist Stuart Hameroff said to me "I don't think an electron has consciousness but I think it has some kind of proto-consciousness", I felt like I probably disagreed -- but also wondered whether we were just getting tangled up in webs of language. I tend to agree with Charles Pierce, Spinoza, Galen Strawson and others that drawing a rigid distinction between mind and matter is ultimately logically incoherent.

I find Cosmism an agreeable attitude toward myself and the universe, but I'm also aware of its limitations. I can't help questing toward a more fundamental understanding of the Cosmos, even though I suspect my limited human brain isn't going to be able to understand the Cosmos too well, and that AGIs, brain enhancement and the like will give us much deeper and better perspectives. I'm currently somewhat bemused by modeling the universe as a process of self-questioning, but unsure how far I'll be able to take the idea.

Often I find myself holding back the part of my mind that wants to spend a lot of time theorizing about such things, because I feel it's more important to work on building AGI! Understanding the world is important to me, but the choice is between trying to understand it directly using my human brain, or building an AGI mind that will be able to help me understand it far better than my human brain is capable of!

One thing that struck my mind when writing this chapter, was the old saw: ***People can adapt to almost anything***... It occurred to me that it makes sense to generalize this to: ***Robust conceptual frameworks can adapt to almost anything***. For example, perspectives like Mormonism and Buddhism and panpsychism can adapt themselves to make meaningful statements about AGI and Singularity, notions that were completely unanticipated (and would have been largely incomprehensible) at the time these religions and philosophies were founded. This says a lot about the adaptable nature of human mind and culture (which emanates, of course, from the adaptable nature of the biological world from which these emerged).

Ultimately, it seems, human conceptual and belief systems are able to adapt to all sorts of new realities, including the Internet and mobile phones and birth control, and soon including AGI and molecular assemblers and cyborgs and what-not. Spirituality and religion embody key aspects of human nature, such as the quest for deep fundamental understanding of the universe, and deep communion with the universe and with other minds – and these quests will keep on going as technology advances, not so much colliding with scientific and technological growth as synergizing with it. Technological revolution will foster a continuous unfolding and expansion of spiritual understanding and experience, which may end up taking a tremendous diversity of different forms – far stranger (and perhaps far deeper) than wild ride of ideas we've rolled through in this chapter.

And so, as I already emphasized, I can make one statement about the contents of this chapter with great confidence: ***Every single idea posited here will appear to us rather silly and limited***, once we have expanded our world view via intense communication and possible fusion with trans-human AGI systems. In other words: Where the cosmos is concerned, we humans don't understand much! We understand more than non-human animals do, and more than pre-civilized humans did... But the AGIs we create will understand much, much more. Which is not to say that any mind will ever come to a complete and thorough understanding of intelligence and the cosmos – maybe it will, maybe it won't; that's among the many things humanity currently has no clue about.

Just as the contradiction between quantum theory and general relativity (gravity theory) tells us that current physics MUST be seriously incomplete in some ways ... so, I feel, does the human race's wild diversity of confusingly contradictory and complementary views on the cosmos tell us that we really don't understand this world we're living in, and our intelligence's place in it, and the roles and properties that will be played by AGIs we create. We don't necessarily understand what our current activities are going to lead to, any better than the "cavemen" who first created language understood that their invention was going to lead to Dostoevsky, differential calculus, Prolog, Google Talk and "Neuroscience for Dummies."

The various concepts and perspectives we're currently experimenting with (and in some cases pouring our hearts and minds into) – classical and quantum information theory, hypercomputing, implicate orders, Mormonism and Buddhism and Cosmism and what-not – all exist within the limited scope of current human individual and collective mind. These perspectives may help us cope with the changes that we are now wreaking in ourselves and the world, via creating AGI and a host of other technologies; and they may well affect the particular nature of the technologies we create and the future world these technologies help create. But these perspectives will then be subverted, and made to seem quaint and ridiculous, by the greater intelligence to which they will lead. And yet, we have to work on refining and elaborating our current perspectives on the world – knowing they will seem absurd and limited in hindsight, from the perspective of greater intelligence – because this refinement and elaboration is part of the process of bringing about said greater intelligence.

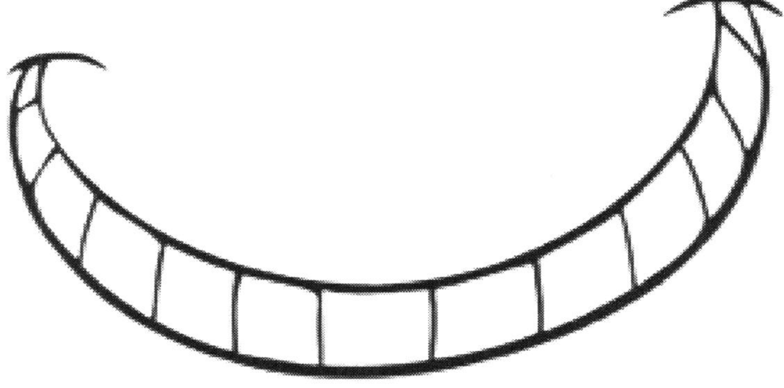

Made in the USA
Lexington, KY
18 January 2015